RAND NATIONAL DEFENSE RESEARCH INSTITUTE

T0108788

Right-Sizing Marine Corps Intermediate Supply Units

A Staffing Model and Proposed Process Improvements

Joslyn Fleming, Ellen M. Pint, Jessica Duke, Simon Véronneau, Austin Lewis, Jordan R. Reimer

For more information on this publication, visit www.rand.org/t/RR2908

Library of Congress Cataloging-in-Publication Data is available for this publication.
ISBN: 978-1-9774-0270-7

Published by the RAND Corporation, Santa Monica, Calif.
© Copyright 2019 RAND Corporation
RAND® is a registered trademark.

Cover: U.S. Marine Corps photo by Cpl. Alexander Mitchell/released.

www.rand.org

Preface

In 2015, intermediate-level supply accounts, such as the Supply Management Unit (SMU) and Repairable Issue Point (RIP), experienced a reduction in personnel as a result of Marine Corps force reductions.[1] This downsizing of supply units, coupled with the increased demand for distributed forces requiring supply support, has created a stress on the supply force. Intermediate-level supply personnel have voiced concerns about not being able to adequately support Marine Expeditionary Forces (MEFs) due to reduced manpower and increased requirements to support exercises and contingencies. However, intermediate-level accounts are currently unable to quantify the impact of the reduction in capacity and the effect on MEF readiness.

This report presents a methodology that can be used to determine capacity for Marine Corps intermediate-level supply and determine the recommended personnel numbers depending on the supported force. This will help the Marine Corps assess whether intermediate supply accounts are manned sufficiently to meet the supported units' requirements. We also propose process improvements that could help intermediate supply units work more efficiently in garrison.

This research was sponsored by the United States Marine Corps Operational Analysis Directorate and conducted within the Acquisi-

---

[1] Per U.S. Marine Corps, Marine Corps Order (MCO) 4400.201, *Management of Property in Possession of the Marine Corps*, Vol. 3, Washington, D.C.: Headquarters Marine Corps, June 13, 2016. Intermediate-level supply accounts typically exist between the consumer and wholesale levels of inventory and support a defined geographic area or provide tailored support to specific organizations or activities.

tion and Technology Policy Center of the RAND National Defense Research Institute, a federally funded research and development center sponsored by the Office of the Secretary of Defense, the Joint Staff, the Unified Combatant Commands, the Navy, the Marine Corps, the defense agencies, and the defense intelligence community.

For more information on the RAND Acquisition and Technology Policy Center, see www.rand.org/nsrd/ndri/centers/atp or contact the director (contact information is provided on the webpage).

Contents

Figures and Tables

Figures

Tables

Summary

The United States Marine Corps (USMC) has faced repeated force reductions in the past decade due to budget cuts, sequestration, and redeployments from operations in Iraq and Afghanistan. USMC logistics units were not spared, and in 2015 intermediate-level supply accounts experienced a reduction in personnel authorizations.[2] At the same time, a movement toward supporting smaller-sized units in more distributed operations places a heavy burden on logistics support elements. Under the dual impositions of reduced manpower and increased requirements, intermediate-level supply forces may not be able to meet the needs of future contingency operations, and further personnel reductions may place additional strain on intermediate supply capacity.

The purpose of this study was to determine if intermediate-level supply units are manned sufficiently to meet their supported units' needs. The USMC must be able to quantify the effects of personnel reductions on intermediate-level supply capacity to determine necessary capacity under possible future scenarios. Accordingly, the research objectives were to (a) create a capacity model for USMC intermediate-level supply accounts and determine personnel requirements based on the size of a supported force; (b) use the capacity model to define a personnel manning formula to inform staffing decisions for intermediate accounts; and (c) to document challenges and best practices experienced by intermediate supply units. To achieve these objectives, RAND

---

[2] Based on J. R. Wilson, "Marine Corps Update: The Frugal Force Faces More Cuts," *Defense Media Network*, 2014, and data from the Total Force Structure Management System (see Chapter 1).

researchers conducted site visits to observe intermediate-level supply processes and document them in a business process map, created a capacity model that assesses intermediate-level supply capacity, and examined the effects of workload drivers, such as the number of units supported, on the number of daily transactions. The research team also reviewed Marine Corps documents and other studies related to intermediate-level supply policies and processes, including service doctrine, standard operating procedures (SOPs), and extant business process maps.

Intermediate Ground Supply in the United States Marine Corps

Our research focused on the two main types of intermediate-level supply organizations in each Marine Expeditionary Force (MEF): the Supply Management Unit (SMU) and the Repairable Issue Point (RIP). The SMU is responsible for managing intermediate-level inventories of consumables in support of a defined geographic area or a specific set of organizations or activities. It provides an intermediate level of inventory between the wholesale level of supply and consumer-level inventory, which is limited in range and depth and intended for internal consumption by the unit. The RIP is responsible for the management of secondary depot repairable and field-level repairable items. Though it is part of the MEF's maintenance battalion and thus a separate organization from the SMU, it serves a similar intermediate function as a buffer between customers in units and the repair facilities that support them at the intermediate and wholesale levels.

To document intermediate-level supply processes, the research team conducted field visits to the SMUs and RIPs at each of the three MEFs. While we found broad similarities in processes across the MEFs, as well as some common challenges, we also found unique challenges at each MEF, as well as best practices that could be shared with the other MEFs.

Subject matter experts (SMEs) we interviewed indicated that intermediate supply organizations face challenges related to the availability of personnel, particularly mid-level supervisors in pay grades

E-5 to E-7. Materiel handling equipment, such as forklifts, pallet jacks, and conveyor systems, is often insufficient, inoperable, or outdated, which increases the amount of manual labor and the time needed to complete basic tasks. At some of the MEFs, SMEs reported that financial management personnel require the SMUs and RIPs to slow down or speed up their spending to smooth out fluctuations in spending patterns in other parts of the MEF. As a result, they are not always able to replenish inventories in a timely way to support using units.

Some of the best practices we found were the use of International Organization for Standardization (ISO) certification and Lean Six Sigma methods to codify and improve processes at the I MEF SMU; delegation of budget authority to the SMU and RIP at II MEF; and a supply capacity study conducted at III MEF to measure daily workload and availability of personnel.

Workload Demand Model

To determine if USMC intermediate-level supply units are sufficiently manned to meet their supported units' needs, we created a workforce demand model that estimates the average number of hours worked per day per Marine on intermediate supply tasks. We also analyzed the relationship between demand and number of supported units to better understand whether the current structure is sufficient to support more distributed operations.

Based on a review of different types of workload models and the availability of data on workload, the number of personnel present in units, and productivity in intermediate supply organizations, we modeled the capacity of the SMU section that performs storage tasks using a task-frequency-duration model.[3] The sequencing of tasks (specifically: receiving, stowing, picking, packing, shipping, and inventory)

---

[3] The storage section is the largest section in an intermediate supply unit and has historically experienced the largest cuts in personnel; therefore, it was the primary focus of our modeling efforts. Sufficient data were not available to estimate similar models for the other sections of the SMU and the RIP. However, the same methodology could be applied if the data were available.

and their associated frequency and duration were used to estimate the total number of work hours per day, which we divided by the number of Marines on hand to perform storage tasks. The model yielded a daily estimate of the number of hours worked per Marine, based on the number of tasks completed each day, the average duration of each task, and the number of Marines on hand to perform storage tasks.

We found that the estimated average daily hours in the SMU storage section have risen over the past four years. At III MEF, average daily work hours rose from about three hours in fiscal year (FY) 2013 to a little over seven hours in FY 2018, with a visible jump immediately following the FY 2015 force reduction.[4] However, in practice Marines are expected to spend only five to six hours per day on storage tasks, because of other duties such as annual training, weapons training, enlisted professional military education, and skills training in their military occupational specialty (MOS), as well as to allow for surge capacity when demand is highest.[5] Our model suggests that not only did the average daily hours exceed the ideal number in FY 2018, but on 36 percent of workdays, storage Marines spent more than eight hours on storage tasks alone. To reduce average daily work hours to six, the average number of enlisted personnel on hand in the storage section at III MEF would need to increase from 36 to 42, or about 22 percent.[6]

Recommendations

Based on our examination of intermediate supply processes in the three MEFs and the results of our model, we make the recommendations noted below. Our recommendations for business process improvement

---

[4] Due to good data availability from III MEF, particularly time durations for tasks, it was the focus of our modeling efforts.

[5] Estimate provided by SMEs and study advisory group.

[6] These findings are specific to III MEF. Similar modeling was run for I and II MEF, but lacked time data for those specific units. We used III MEF time data to provide a rough estimate, but due to unique aspects of III MEF storage operations, such as warehouse design and its ability to pull workload twice a day, we limited our model findings in this report to III MEF.

are based on our observations and interviews during site visits; additional analysis may be needed to measure costs and benefits.

Right-Sizing Intermediate Supply Units

- **Increase authorizations or staffing goals for 3051s:** Based on modeling of the III MEF storage section, to reduce average daily work hours to six, the number of enlisted personnel on hand would need to increase by 8–17. This can be achieved by either an increase in authorizations or a higher staffing goal. Additional efficiencies can be gained in garrison from process efficiencies, but our primary recommendation is to increase staffing to account for deployed operations where garrison efficiencies will not have impact.

Improving Staffing Modeling

- **Archive on-hand personnel data:** Intermediate supply capacity could be better measured with accurate data on the number of Marines available in each section. The Marine Corps should develop a process to archive personnel counts that are recorded daily in Marine Online.
- **Collect data on task frequency and duration for all SMU and RIP sections to improve capacity models:** To develop task-frequency-duration models similar to the one we created for the storage section, additional data are needed on the types of tasks performed by each section, the frequency of each task, and the average amount of time to complete the task.
- **Forecast staffing requirements for deployed environments:** Further data collection and analysis of supply support for deployed organizations is needed to forecast the number of intermediate supply personnel required based on the characteristics of the supported units.

Improving Business Processes

- **Provide basic material handling equipment (MHE) found in all U.S. commercial warehouses:** We observed a lack of basic material handling equipment at the MEFs, which affects the efficiency of operations and the potential for workplace injuries. Providing improved equipment could improve productivity and reduce the number of Marines needed to perform daily workload in garrison.

- **Ensure that Marines have the correct personal protective equipment (PPE) to perform their work:** Marines did not appear to have access to appropriate safety boots and shoes, eye protection, high-visibility vests, and ear protection. Given the small workforce, workplace safety should be a higher priority to avoid any lost days caused by accidents.

- **Allow the SMUs and RIPs at each MEF to control their spending rate:** The RIP and SMU should be free to alter their spending rate over the fiscal year in response to customer demands, while staying within total budgetary bounds. The demand signals for consumables and repairables are not steady throughout the year because of deployments, exercises, and other changes in unit activities.

- **Item locations should be further assessed and optimized at both RIPs and SMUs to reduce picking time:** A significant reduction in transaction times at RIPs and picking time at SMUs could be achieved by rearranging items so that the fast-moving items are as close as possible to the service counters in the case of RIPs and to the shipping and receiving areas at SMUs. Further assessment is recommended.

- **Foster a Lean Six Sigma (LSS) Program for Continuous Improvement:** The LSS program at I MEF is a source of pride for SMU Marines and has helped other organizations on Camp Pendleton improve their processes. With leadership support, similar initiatives could be implemented at the other MEFs.

Acknowledgments

The authors wish to acknowledge the invaluable support provided by our sponsors, Master Gunnery Sergeants Brady Wentlandt and Frederic Zeyer from Marine Corps Installations and Logistics, and our study monitor, Major Jason Fincher of the Marine Corps Operations Analysis Directorate. They provided tremendous assistance throughout this study.

We are also grateful to the other members of the Study Advisory Committee and other stakeholders for their input and guidance. We wish to thank the supply personnel of I, II, and III MEF who welcomed us into their offices and warehouses and provided so much of their time to help support this project. Thank you for your hospitality and support.

We thank our RAND colleagues Chris Mouton and Mike Decker and our dedicated reviewer, Ken Girardini, for providing helpful advice and feedback on this research, as well as James Broyles for his expertise and input throughout the course of the study.

Abbreviations

3PL	third-party logistics
CCI	controlled cryptographic item
Class IX	repair parts
CLB	Combat Logistics Battalion
CWO	chief warrant officer
DASF	Due and Status File
DLA	Defense Logistics Agency
DRMO	Defense Reutilization Marketing Office
DSU	Deployed Support Unit
ELMACO	Electronics Maintenance Company
FY	fiscal year
GA	General Account
GCSS-MC	Global Combat Support System–Marine Corps
GS CLR	General Support Combat Logistics Regiment
HAZMAT	hazardous material
HQMC	Headquarters Marine Corps
IIP	Initial Issue Provisioning
IMA	Intermediate Maintenance Activity
IMO	Item Master Organization
ISO	International Organization for Standardization
IT	information technology

LSS	Lean Six Sigma
MAGTF	Marine Air-Ground Task Force
MCB	Marine Corps base
MCO	Marine Corps order
MCSS	Marine Corps Supply System
MEF	Marine Expeditionary Force
MEU	Marine Expeditionary Unit
MHE	material handling equipment
MMDC	MAGTF Materiel Distribution Center
MOS	military occupational specialty
MVGL	Money Value Gained Lost
NCO	noncommissioned officer
NIIN	National Item Identification Number
NIPRNet	nonclassified internet protocol router network
P2	preservation and packaging
P3	packing, packaging, and preservation
PME	professional military education
PPE	personal protective equipment
RFA	request for assistance
RIP	Repairable Issue Point
RMC	Repairable Maintenance Company
SASSY	Supported Activities Supply System
SECREP	secondary reparable
SME	subject matter expert
SMU	Supply Management Unit
SOP	standard operating procedure
SPMAGTF	Special Purpose Marine Air-Ground Task Force
STRATIS	Storage Retrieval Automated Tracking Integrated System

TEEP	Training Exercise and Employment Plan
T&R	training and readiness
T/O	table of organization
USMC	United States Marine Corps

CHAPTER ONE

Introduction

The United States Marine Corps (USMC), already the smallest U.S. military service and the "frugal force"—to quote recent Commandant Gen. James Amos—has faced repeated force reductions in the past decade due to budget cuts, sequestration, and redeployments from operations in Iraq and Afghanistan.[1] USMC logistics units were not spared, and in 2012 intermediate-level supply accounts experienced a reduction in personnel as a result of these policy changes.

And yet, as stated in the most recent Marine Corps Operating Concept:

> Our logistics enterprise has to provide expeditionary support and sustainment from the greater distances imposed by A2AD [anti-access/area denial] threats. . . . It must be capable of supporting distributed units on a widely dispersed battlefield and reduce the burden that must be carried by the individual Marine. . . . We must . . . rebalance our logistics capabilities between the whole-sale/bulk level and the retail/individual level, reducing the traditional logistics stockpile ashore and relying on a transportation/distribution system that delivers [supplies] to smaller, dispersed units . . . across a large geographic area.[2]

---

[1] J. R. Wilson, "Marine Corps Update: The Frugal Force Faces More Cuts," *Defense Media Network*, 2014.

[2] U.S. Marine Corps, *Marine Corps Operating Concept: How an Expeditionary Force Operates in the 21st Century*, Washington, D.C.: Headquarters Marine Corps, September 2016, p. 23.

This call for support to smaller-sized units and more distributed operations places a heavy burden on logistics support elements. Though the Marine Corps prides itself for "doing more with less," leadership is nevertheless concerned about the Marine Corps' ability to meet future demand under the dual impositions of reduced manpower and increased requirements.[3] While current intermediate-level supply forces may be able to meet current needs in garrison, the amount of workload may be creating stress on the force, and possible future contingency operations and/or further personnel reductions may place additional strain on logistics capacity.

To determine necessary capacity under possible future scenarios, the Marine Corps must first be able to quantify the effects of personnel reductions on intermediate-level supply capacity. During the October 2016 Ground Supply Training and Readiness Conference, USMC supply subject matter experts (SMEs) expressed an interest in defining supply capacity and relating it to MEF readiness and manpower strength. Additionally, a May 2016 RAND report, *Developing a Capacity Assessment Framework for Marine Logistics Groups*, developed a framework to assess logistics capacity, but noted the difficulty in understanding and measuring supply capacity at the "general support" (i.e., intermediate) level.[4] Therefore, the study documented in this report undertook a more concentrated effort to develop a methodology to measure supply capacity specifically at the intermediate level.

Research Objectives and Tasks

The purpose of this study was to determine if intermediate-level supply units are manned sufficiently to meet their supported units' needs.

---

[3] Raymond Priest, "Doing More with Less," *Marine Corps Gazette*, Vol. 74, No. 10, 1990.

[4] Joslyn Hemler, Yuna H. Wong, Walter L. Perry, and Austin Lewis, *Developing a Capacity Assessment Framework for Marine Logistics Groups*, Santa Monica, Calif.: RAND Corporation, RR-1572-USMC, 2017.

Accordingly, the research objectives were to (a) create a capacity model for USMC intermediate-level supply accounts and determine personnel requirements based on the size of a supported force; (b) use the capacity model to generate staffing requirements for intermediate accounts; and (c) to document challenges and best practices experienced by intermediate supply units. RAND researchers undertook three tasks to achieve these objectives:

- document current intermediate-level supply processes in a business process map
- create a capacity model that assesses intermediate-level supply capacity and is adaptable to any unit
- use the model to estimate staffing requirements based on the expected level of demand from the supported unit.

The research team reviewed Marine Corps documents and other studies related to intermediate-level supply policies and processes, including service doctrine, standard operating procedures (SOPs), and extant business process maps. It also analyzed USMC data from a variety of sources: reviews of key supply tasks (e.g., unit mission essential task lists [METLs]), manning documents (e.g., tables of organization and equipment [T/O&Es]), and historical supply data (e.g., requisition histories in Global Combat Support System–Marine Corps [GCSS-MC] and workload metrics in the Storage Retrieval Automated Tracking Integrated System [STRATIS]). It then conducted semistructured interviews with key stakeholders during site visits to Supply Management Units (SMUs) and Repairable Issue Points (RIPs) at each of the three Marine Expeditionary Forces (MEFs). These efforts resulted in the identification of intermediate-level supply tasks, demand for those tasks, current staffing and resourcing, and metrics associated with each task. Research team members also updated business process maps, identified best practices that could be shared among the MEFs, and proposed process improvements that could help intermediate supply units work more efficiently in garrison.

Based on this output, the team created a model that assesses intermediate-level supply capacity and the ability to meet supported units' needs. After reviewing different types of workforce capacity models and the available data for intermediate supply units, we selected a task-frequency-duration workforce demand model. This allowed us to calculate the average number of hours worked per day per Marine on intermediate supply tasks, and given target available hours per day (e.g., six hours), to estimate the total number of Marines required. We then compared the results to current staffing to determine whether any additional personnel were needed to fulfill current requirements. We also analyzed the relationship between demand and number of supported units to better understand whether the current structure is sufficient to support more distributed operations.

Intermediate Ground Supply in the United States Marine Corps

The Marine Air-Ground Task Force (MAGTF) is the USMC's principal organization for all missions across the range of military operations. A MAGTF is a balanced air-ground, combined arms task organization under a single commander and is structured to accomplish a specific mission. Each MAGTF contains four core elements: a command element, a ground combat element, an aviation combat element, and a logistics combat element.[5] Within the logistics combat element, the Marine Corps Supply System (MCSS) is tasked with providing the MAGTF commander with the necessary materiel for conducting combined arms operations.

---

[5] The MEFs are the Marine Corps' primary "standing MAGTFs" in peacetime and wartime. A Marine Expeditionary Brigade is a scalable MAGTF with a force of up to 20,000 Marines and Sailors. A Marine Expeditionary Unit (MEU) is organized as a MAGTF with approximately 2,600 Marines and Sailors and provides a geographic combatant commander with a rapid response force either independently or as part of a larger amphibious, joint, or combined force. If a MEU is inappropriate or unavailable for a specific mission, a Special Purpose MAGFT (SPMAGTF) is formed. A SPMAGTF is normally no larger than a MEU, with tailored capabilities required for accomplishing a particular mission. See U.S. Marine Corps, "Types of MAGTFs," *Marine Corps Concepts and Programs*, Washington, D.C.: Headquarters Marine Corps (HQMC), 2017.

The MCSS consists of three managerial levels: Headquarters Marine Corps (HQMC) and the in-stores and out-of-stores functional elements. The first two comprise the wholesale inventory level of supply, overseeing the performance of the entire supply system and serving as an initial distribution point, respectively. The third managerial level deals with retail inventory, which includes both intermediate and consumer inventory (see Figure 1.1).[6] Per USMC doctrine, intermediate-level supply stands between the "consumer- and wholesale-levels of inventory for support of a defined geographic area or tailored support of specific organizations or activities."[7] Its roles and procedures are currently governed by Marine Corps Order (MCO) P4400.151B, Ch 2, which establishes "supply policies necessary in the effective control of intermediate-level inventories."[8]

Figure 1.1
The Marine Corps Supply System

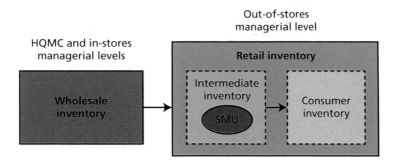

SOURCE: Abercrombie, Fullbright, and Long, 2016.

---

[6]   Erick Abercrombie, Patrick Fullbright, and Seth Long. "Analysis of the Marine Corps Supply Management Unit's Internal Operations and Effect on the Warfighter," MBA Professional Report, Monterey, Calif.: Naval Postgraduate School, 2016.

[7]   U.S. Marine Corps, Marine Corps Order [MCO] P4400.151B, Ch 2, *Intermediate-Level Supply Management Policy Manual*, Washington, D.C.: Headquarters Marine Corps, December 14, 2012, p. v. Henceforth cited as MCO P4400.151B, Ch 2.

[8]   MCO P4400.151B, Ch 2, p. v. Note that this project focuses on ground supply—i.e., in support of the MAGTF ground combat element; it does not include aviation supply, which supports the air combat element.

Supply Management Units

Supply Management Units are responsible for managing intermediate-level inventory in support of a defined geographic area or a specific set of organizations or activities. Given its position, the SMU serves as a buffer between the wholesale-level of supply and consumer-level inventory, which is limited in range and depth and intended for internal consumption by the unit.[9] Accordingly, the degree to which the SMU can effectively stockpile supplies and fulfill and ship requisitions has a "direct impact on the supply and maintenance readiness levels of its supported units."[10] SMUs are generally stationary and do not typically deploy with units during operations overseas, but may provide personnel for the logistics combat element of a Marine Expeditionary Unit (MEU) or other type of MAGTF.[11]

Each MEF has its own SMU, located at its respective headquarters: Marine Corps Base (MCB) Camp Pendleton in California (I MEF), MCB Camp Lejeune in North Carolina (II MEF), and MCB Camp Butler in Okinawa, Japan (III MEF). A SMU consists of several sections that are responsible for various supply support tasks, which are summarized in Table 1.1. For example, the General Account (GA) section is responsible for stock control, daily reports, reconciling the Due and Status File (DASF), and managing special projects, such as disposition services, shipping confirmations, and analyzing Money Value Gained Lost (MVGL) reports. The customer service section is responsible for handling requests for assistance (RFAs) from supported units and fulfilling walk-through requisitions, which allow units to pick up mission-critical supplies without having to wait for the routine picking, packing, and shipping processes. The storage section is the largest section in the SMU, and is responsible for receiving deliveries from wholesale inventory, stowing repair parts (Class IX) and other consumable items, picking and verification of items ordered by

---

[9]   MCO P4400.151B Ch 2, p. v.

[10] Abercrombie, Fullbright, and Long, 2016, p. 1.

[11] U.S. Marine Corps, *MAGTF Supply Operations*, Marine Corps Tactical Publication 3-40H, Washington, D.C.: Headquarters Marine Corps, February 19, 1996, updated through May 2, 2016, p. 3-3.

Table 1.1
Supply Management Unit Sections and Tasks

Section	Tasks
General Account	Stock control
	DASF management
	Special projects
Fiscal/Procurement	Stock control buying
Customer Service	Reconciliation
	Walk-through
	Shipping discrepancy
	GCSS-MC support
	Training and readiness (T&R) support
Deployment Support Unit	Forward deployed unit support
	Build/manage Class IX blocks
Initial Issue Provisioning	Receipt
	Issue
	Storage
Storage/Warehousing	Ship/receive
	Stowing
	Picking
	Verification
	Inventory
Packing	Pack
	Outbound receiving
	Hazardous material (HAZMAT) processing
	Box building
MAGTF Materiel Distribution Center	Receive
	Distribute
	Track and trace

SOURCE: Total Force Structure Management System, Abercrombie, Fullbright, and Long, 2016.

units, and conducting physical inventories of items held in the SMU warehouse.

The number of authorized enlisted personnel in each SMU has been reduced over the last five years as the Marine Corps downsized

Figure 1.2
Reduction in Supply Management Unit Enlisted Personnel at I Marine Expeditionary Force, FY 2013–2018

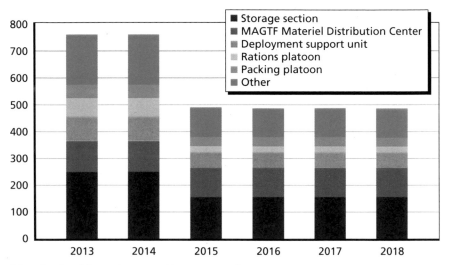

SOURCE: Total Force Structure Management System.

from a peak of 202,000 personnel to a force of approximately 186,000. For example, Figure 1.2 shows the number of enlisted authorizations in each section of the 1st Supply Battalion at I MEF, based on the unit Table of Organization (T/O) in January of each year. The most dramatic reductions occurred in 2015, when total enlisted authorizations fell from 758 to 485. Similar cuts were experienced by the SMUs in the other MEFs. The storage section experienced the largest reduction in authorized personnel of almost 100 billets. Not only were most sections reduced in staff, but three sections with 63 authorized enlisted personnel were eliminated. These included the sections providing augmentation to the General Support Combat Logistics Regiment (GS CLR), and the MEU Combat Logistics Battalion (CLB), and the Supported Activities Supply System (SASSY) management unit.[12]

---

[12] SASSY was the logistics information management system that has been replaced by GCSS-MC.

Table 1.2
Changes in Supply Management Unit Enlisted Personnel at I Marine Expeditionary Force, FY 2013–2018

Section	FY 2013 Enlisted Personnel	FY 2018 Enlisted Personnel
Storage	251	158
MAGTF Materiel Distribution Center	114	110
Deployment Support Unit	93	57
General Account	32	35
Packing Platoon	50	34
Inventory Control Unit	30	23
Headquarters Operations	22	21
Rations Platoon	65	20
Customer Service	18	13
Headquarters Platoon	8	8
Fiscal	12	6
Supply Co Augments (GS CLR)	39	0
MEU CLB Plus-Up Supply	19	0
SASSY Management Unit	5	0
Total	758	485

SOURCE: Total Force Structure Management System.

While these sections were eliminated, SMU personnel we interviewed indicated that similar functions are still being performed by other sections. Changes in the number of enlisted personnel in I MEF's T/O for each SMU section from fiscal year (FY) 2013 to FY 2018 are summarized in Table 1.2.

Repairable Issue Point
The RIP is responsible for the management of secondary repairable and field-level repairable items. Though it is a section of the MEF's main-

tenance battalion and thus not part of the SMU, it is included with intermediate-level supply organizations in MCO P4400.151B Ch 2. It conducts direct exchange of unserviceable items from customers for serviceable items from its inventory, sends unserviceable items to intermediate maintenance activities (IMA), depot maintenance activities, or commercial vendors, and receives and distributes items from supply sources and maintenance to fill customer backorders and replenish its own inventories. The RIP budgets and manages funds for asset replenishment and is responsible for maintaining adequate inventories to support customers.[13] Sub-RIPs are normally established for geographically separated customers and deployed units.

The tasks performed by the RIP are summarized in Table 1.3.

Like the SMUs, the RIPs experienced significant reductions in authorized enlisted personnel in 2015. Figure 1.3 shows changes in the RIP T/O for II MEF, including changes in the two primary military occupational specialties (MOSs), Supply Administration and Operations Specialist (3043) and Warehouse Clerk (3051). The RIPs in the other two MEFs had similar reductions in these two primary MOSs, although I MEF had some offsetting increases in other MOSs, primarily motor vehicle and wrecker operators who perform functions outside the RIP warehouse.

Table 1.3
Repairable Issue Point Tasks

Section	Tasks
Repairable Issue Point	Returns and issues to/from customer
	Issues and receipts to/from IMA and contractor
	Returns, orders, and receipts to/from wholesale
	DASF management
	Financial management

SOURCE: Based on MCO P4400.151B Ch 2, chapter 5.

---

[13] MCO P4400.151B, Ch 2, chapter 5. Customer units are not charged for repairable items unless they fail to turn in an unserviceable asset.

Figure 1.3
Reduction in Repairable Issue Point Enlisted Personnel at II Marine Expeditionary Force, FY 2013–2018

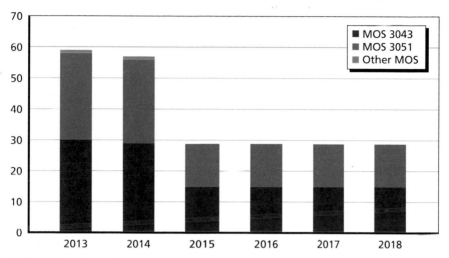

SOURCE: Total Force Structure Management System.

Increased Demand for Intermediate Supply Functions

The reductions in SMU and RIP personnel have led to concerns about their ability to adequately support their respective MEFs. Combined with these reductions, intermediate supply accounts have seen an increase in requirements to support exercises and contingency operations, putting further stress on their ability to meet customer demand. The Marine Corps has created new force packages, such as Special Purpose Marine Air-Ground Task Forces (SPMAGTFs), to respond to a variety of regional crises.[14] These units, established in each of the geographic combatant commands, require their own intermediate supply capabilities, which are often sourced from the garrison intermediate supply units. This causes further reductions in personnel available to support SMU and RIP operations in garrison and an increase in support of deployed units. Additionally, as the Marine Corps moves to

---

[14] Andrew Feickert, "Marine Corps Drawdown, Force Structure Initiatives, and Roles and Missions: Background and Issues for Congress," Washington, D.C.: Congressional Research Service, CRS Report No. R43355, 2014.

an emphasis on employing battalion- and company-sized units, the number of units requiring intermediate-level support continues to increase. This decentralization and distribution of forces will continue to place increased demand on intermediate supply units.

Other external personnel training demands have levied additional taxes on intermediate supply units and limit the availability of Marines to work directly on supply tasks. Over the past 10–20 years, there has been a large increase in mandatory training requirements across the military services, and the same holds true for the Marine Corps. In 2002, the average Army company commander had 256 days available for training annually, but to accomplish all required training would have taken 297 days, leaving a deficit of 41 days. By 2015, total training requirements had increased to over 500 days, resulting in a deficit of 258 days.[15] Similarly, in Marine Corps intermediate supply units, supply Marines are often taken away from their day-to-day supply jobs to meet requirements such as annual training, weapons training, watch standing, enlisted professional military education (PME), and MOS skills training. SMEs in the MEFs with whom we spoke estimated that over half of their Marines are conducting tasks other than supply-related operations each day.

The combination of all these factors has led to what appear to be possible shortfalls in the manpower needed to provide intermediate-level supply functions. The purpose of this study was to develop a set of models for assessing intermediate supply capacity relative to current demands.

Organization of This Report

The remainder of this report presents the result of our research. Chapter Two provides a description of intermediate supply processes at the SMU and RIP. Chapter Three gives an overview of workload capacity models and details the models used to determine intermediate supply

---

[15] Crispin J. Burke, "No Time, Literally, for All Requirements," Association of the United States Army, 2016.

capacity. Chapter Four includes best practices and areas for improvement. Chapter Five concludes with findings, recommendations, and broader implications for determining future workforce capacity. Appendix A provides greater detail on intermediate supply processes and detailed process maps. Appendix B describes the evolution of the workload capacity model and a Monte Carlo simulation that allows for variation in task times and the number of personnel on hand.

CHAPTER TWO

Supply Management Unit and Repairable Issue Point Processes

This chapter outlines current Marine Corps intermediate ground supply processes and structure. To document these processes, we conducted field visits to the SMU and RIP facilities at each of the MEF's (Camp Pendleton, Camp Lejeune, and Camp Butler).[1] At each site, we followed a field study methodology that included semistructured interviews, observations of work flow, and process analysis. Each visit followed a similar approach. First, we held an initial meeting with unit leadership and key section staff to explain the purpose of the visit and ask MEF-specific questions. Next, we conducted walk-through tours of the various sections of the SMU and the RIP. In each section, we followed a supply chain logic, starting from the point when inbound items arrived and ending when outbound items left the section. The tours and discussions clarified the section's roles and missions. Warehouse tours typically followed the processing of parts from arrival to storage and from retrieval to outbound shipment, while office meetings focused on supporting functions, such as GA and fiscal.

The primary tour guide in each section was the associated chief warrant officer (CWO), company-grade officer, or senior noncommissioned officer (NCO). We also spoke with other NCOs, junior enlisted Marines, government civilians, and contractor personnel to understand the tasks they performed and to obtain supplementary informa-

---

[1] II MEF was visited December 11–15, 2017, followed by I MEF January 22–26, 2018 and III MEF March 12–16, 2018.

tion. These unstructured discussions helped us understand daily challenges, refine our process maps, and clarify issues at hand. We allowed time during the field visit to revisit certain sections if key personnel were absent or we needed to ask follow-up questions. Although our field visits focused primarily on the RIP and SMU, we also visited some maintenance units and customer units to gain a different perspective on the intermediate supply structure. Table 2.1 shows a summary of the number and types of personnel who took part in discussions with us.

Based on our site visits and discussions with intermediate supply SMEs, as well as existing process maps and reference documents, we created general process maps of SMU and RIP functions. During our visits, we also identified best practices and challenges at each MEF. While the field visits demonstrated broad similarities in processes across the MEFs, we noted some differences in operations. In the remainder of this chapter, we summarize SMU and RIP processes by section. Understanding the processes and tasks associated with each section was instrumental in helping to develop the workload capacity models that are discussed in the next chapter. However, we focused the modeling efforts on the storage section, the largest of the SMU and RIP sections.

In Chapter Four, we discuss common challenges across all three MEFs, as well as best practices and unique challenges in each MEF. A description of the general intermediate ground supply process is provided in Appendix A, along with detailed process maps.

Table 2.1
Number of Personnel Who Participated in Discussions

Type of Personnel	I MEF	II MEF	III MEF
Commissioned Officers (O1–O6)	7	4	20
Warrant Officers (W1–W3)	4	5	11
Senior NCOs (E7–E9)	8	5	10
NCOs (E5–E6)	14	7	6
Junior Enlisted (E2–E4)	Multiple	Multiple	Multiple
Government Civilians	2	2	1
Contractor Personnel	3	2	2

Supply Management Unit Processes by Section

The SMU serves as the primary source of Class IX consumable repair parts for ground support. It has two overarching functions: order fulfillment for using units and warehouse management of new consumables. To accomplish these functions, the SMU is separated into sections, primarily along MOS lines, which perform various processes (Table 2.2). The storage section is also further separated into specific subsections: storage, packing, and receiving. The general intermediate ground supply process is described in detail in Appendix A.

General Account Tasks

The GA section ensures that the SMU can maintain an appropriate level of consumables for supported units by determining the necessary stockage level and order frequency based on the items' priority in GCSS-MC and historical use data from STRATIS, the Marine Corps' warehouse

Table 2.2
Supply Management Unit Sections

Section Name	Section Description	Primary MOS
General Account	Maintains stocking level and order frequency	3043
Fiscal	Ensures financial stability	3043
Customer Service	Supports interactions between using unit and SMU	3043
Deployment Support Unit (DSU)	Supports deployed units through Block IX floats	3043
Initial Issue Provisioning	Receives and Issues new equipment to using units	3051
Storage—Receiving	Receives and receipts incoming orders	3051
Storage—Storage	Stores received items, picks orders, and performs inventory counts	3051
Storage—Packing	Packs orders	3051
MAGTF Materiel Distribution Center	Receives and distributes items to using units	3051

SOURCE: Based on MCO P4400.151B Ch 2, chapter 5.

logistics system. Mission changes and the introduction of new items (identified by National Item Identification Numbers [NIINs]) require the GA section to review and revise the SMU's ordering rates frequently.

Orders for items that are not available locally in the SMU warehouse are flagged for further examination by the GA section. If they are confirmed as a necessary purchase, GA submits the order and it is added to the DASF, which GA monitors daily for order updates and delays.

Fiscal Tasks

The fiscal section maintains the SMU's finances and manages the annual (fiscal year) budget provided by the MEF. It must ensure that there is enough funding available to maintain the SMU's inventories, yet also adapt to changes in demand. Like the GA, the fiscal section relies on historical use data for budget forecasting. Before placing orders, especially with unexpected requests, the GA section confirms with the fiscal section that the purchase is within budget. The fiscal section regularly reviews and updates the SMU's budget to provide the most up-to-date financial information for management and oversight.

Customer Service Tasks

The customer service section interacts with using units in garrison on behalf of the SMU and handles RFAs and reconciliations of the units' property books. When an RFA is submitted, a representative from customer service addresses the issue or provides information as to the next course of action. There are several different RFAs that a using unit can submit to customer service.[2] These include:

- Shipping discrepancies: any issues regarding the delivery of a requested consumable
- Walk-through: a request to pick up an item directly from the SMU instead of having it shipped to the unit through normal processes
- GA issues: any issues pertaining to the GA section

---

[2] As indicated in 1st Supply Battalion, *Supply Management Unit Standard Operating Procedures,* SMUP 4400.1, November 16, 2017.

- Item Master Organization (IMO) loads: a request to add a NIIN to the GCSS-MC database
- Stock checks: a request to verify the SMU's on-hand quantity
- Lateral support: a request for the SMU to support a unit not supported by the MEF
- Track and trace: a request to ensure proof of delivery for cargo
- Causative research: any request not covered by the other RFAs
- MOS training: a request for assistance on supply administration and warehousing procedures
- Demand supported items: a request to access stock items that are limited for authorization to support mission requirements.

Additionally, the SMU customer service section conducts reconciliations with the units that it supports. These are carried out approximately once a month. A Marine from the customer service section will review the materiel property currently listed on the using unit supply section's property books. Additionally, as the local SMEs, the customer service Marines also visit units to provide supply support training.

Deployed Support Unit Tasks

Like customer service for units in garrison, the deployed support unit (DSU) provides SMU support to deployed units, largely through assistance with Class IX Block support. A IX Block contains consumable repair parts and is deployed along with the using unit to serve as a stand-in SMU in a deployed environment. The DSU works with the unit's supply officer to make unit-specific modifications to the standard list of NIINs required for a given environment. Once a list is prepared, the DSU checks the warehouse to confirm that the requested NIINs are available. The IX Block components belong to the DSU section of the SMU until the deployed unit uses them. When a unit returns from deployment, it returns unused consumables from its IX block to the SMU.

Storage Tasks

The SMU storage section is divided into three smaller subsections: storage, packing, and receiving. The number of Marines assigned to each subsection varies daily based on the workload generated by STRATIS indicating what items are to be picked, packed, and sent out that day.[3] SMU warehouses are separated by the size of the NIIN into bins (small consumables), G-Lot (larger consumables), and bulk (bulk consumables). With the exception of bins, which utilize electronic carousels to store the items, the warehouses are all operated similarly regardless of their content.

Storage Subsection

The storage subsection locates and picks requested consumables from the warehouse inventory and replenishes inventories by stowing items received by the SMU. Picks are determined by the daily workload lists generated by STRATIS. Warehousemen pull consumables on the list from the storage area, confirm the NIIN and quantity, record the pull, and set the items aside to be transferred to the packing area. If an item is not where STRATIS indicates it should be, the Marine will conduct a search in the surrounding area. Once the daily picks are completed, the Marines stow incoming items transferred from the receiving subsection. A quality check is performed before the item is stored.

In addition to picking and stowing consumables, the storage subsection conducts spot-check inventory counts of on-hand material in the warehouse. STRATIS supplies a random list of NIINs distributed throughout the warehouse for Marines to check to verify their quantity and location. Inventory counts are also automatically generated if the recorded inventory falls below ten.

Packing Subsection

The packing subsection receives items picked by the storage subsection and prepares them for shipment to the units in garrison through the MEF's MAGTF Materiel Distribution Center (MMDC). Hazardous material (HAZMAT) or other equipment with more stringent packing requirements is handled by the packing, packaging, and preservation (P3) section, a separate organization within the supply battalion. After all of the items on the workload list have been packed and are ready for distribution, the Marines in this subsection assist other sections in the warehouse.

---

[3] The exception is III MEF, which pulls one set of customer orders in the morning and another in the afternoon.

Receiving Subsection

The receiving subsection receives and receipts orders and deliveries coming into the SMU. Since incoming packages may arrive throughout the day, the number of Marines assigned to this subsection fluctuates based on workload. When packages arrive at the SMU, Marines compare the NIINs and quantities in the package with the shipment's manifest. They separate the NIINs according to their warehouse location and transfer them to the storage subsection to be stowed. Items that are either not identifiable or do not match the manifest are set aside for further investigation.

Repairable Issue Point Processes by Section

The RIP's primary function is to receive unserviceable (Code F) secondary reparable (SECREP) items from using units and to replace them with serviceable (Code A) items of the same type. Like the SMU, the RIP is divided into sections based on MOSs; however, because of the RIP's smaller size, its tasks are consolidated into only two sections, Administration and Warehouse (Table 2.3).

Administration Tasks

The administration section handles the operational aspects of the RIP, including customer service interactions, financial operations, and inventory management. This section sets inventory levels and order frequency for SECREPs based on historical use data. Supply Marines

Table 2.3
Repairable Issue Point Sections

Section Name	Section Description	Primary MOS
Administration	Maintains stocking level and order frequency Ensures financial stability Manages backorders Runs front desk and customer service interactions	3043
Warehouse	Receives Code A and Code F Materiel Performs inventory count	3051

SOURCE: Based on MCO P4400.151B Ch 2, chapter 5.

operate the front desk and receive Code F SECREPs from using units. In most instances, they conduct a visual check to confirm that the SECREP is properly configured and determine whether a Code A SECREP of the same NIIN is available as a replacement. If not, the RIP will place the item on backorder to be picked up by the unit at a later time.

When a Code F SECREP is received by the RIP, there are four different destinations where the item can be sent: a Marine Corps IMA, a third-party logistics (3PL) vendor, a wholesale source of supply for repair, or the Defense Reutilization Marketing Office (DRMO) for disposal. The source of repair depends on capability and capacity, as well as the repair cycle time and cost of each source. The administration section's CWO handles the financial aspects of the RIP, along with a small number of civilians. Like the SMU, the RIP must replenish inventories within its annual (fiscal year) budget provided by the MEF.

For deployed units, the administrative section assists in assembling a subfloat of SECREPs that deploys with the unit and works with the Marines in the field to keep the subfloat fully functional.

Warehouse Tasks

Warehousemen perform the storage and inventory functions of the RIP warehouse. They receive both Code A and Code F items returned by maintainers and outside vendors, identify where Code A SECREPs need to be stored, and stow them in the warehouse. Warehouse Marines also perform inventory counts, including spot checks and annual wall-to-wall checks. Discrepancies are brought to the administrative section for further investigation.

Conclusion

The SMU and RIP processes and tasks outlined in this chapter were the foundation for developing workload capacity models that will be presented in the next chapter. Understanding each section's discrete tasks allowed us to build a task-frequency-duration model. Due to limited data availability, we focused our efforts on modeling the largest of the intermediate supply sections, the storage section. While this chapter presents the common tasks across all three MEFs, we further elaborate on MEF-specific challenges and best practices in Chapter Four and provided detailed process maps of these processes in Appendix A.

CHAPTER THREE

Workload Demand Models: Intermediate Supply Functions

To determine if USMC intermediate-level supply units are sufficiently manned to meet their supported units' needs, we created workforce demand models that estimate the average number of hours worked per day per Marine on intermediate supply tasks. We also analyzed the relationship between demand and number of supported units to better understand whether the current structure is sufficient to support more distributed operations.

Considering there are numerous possible approaches to developing a workload demand model, we first summarize a few of the more relevant approaches in this chapter, discussing the advantages and disadvantages of each and describing our rationale behind selecting a particular approach. We then describe how we applied this approach to develop a demand model for the storage section. Because the storage section is the largest section within intermediate supply units, we focused our modeling efforts on that section, where they will have most impact. Due to data and time constraints, the other intermediate supply sections were not modeled to the same extent as the storage section. However, the final section of this chapter will outline recommended model inputs and data sources for other sections.

Workload Demand Model Types

As outlined in a recent RAND study of *Options for Department of Defense Total Workforce Supply and Demand Analysis* (2014), there are

several different types of approaches to determine workforce demand—
some quantitative, others qualitative, and some combining aspects of
each. This section will outline factors common to all workforce demand
models and highlight the strengths and weaknesses of each.

Factors Common to Workforce Demand Models

Adequately projecting future workforce demand generally requires a
three-step process. The first step is to assess current workforce demand,
considering total workload, individual worker productivity, and total
workforce size. This allows one to examine the relationship between
workload and necessary workforce—most simply by dividing the total
work by a measure of worker productivity, which yields an estimate of
workers needed to accomplish the workload.

The next component is to project future workload, considering a
variety of demand drivers that would affect that assessment. These can
include the overall business environment, the organization's strategic
goals and plans, estimated future demand for products or services, projected budgets, and potential opportunities for business development.
Both quantitative and qualitative methods may be appropriate for this
projection, depending on the organization and type of work.

The third step combines the first two, projecting the worker productivity derived in step one into the future and applying it to the future
workload derived in step two, yielding a projection of future workforce
requirements. Naturally, this includes a degree of assumption about
how applicable current or historic information is to future needs. How
to do that with a measure of methodological rigor, whether quantitative, qualitative, or both, is explored in the remainder of this section.[1]

Hybrid Quantitative-Qualitative Approaches

Given the range of strengths and weaknesses of individual methods,
as well as the general limitations of either quantitative (e.g., availability of required data) or qualitative (e.g., finding and convening

---

[1]   Shanthi Nataraj, Christopher Guo, Philip Hall-Partyka, Susan M. Gates, and
Douglas Yeung, *Options for Department of Defense Total Workforce Supply and
Demand Analysis: Potential Approaches and Available Data Sources*, Santa Monica,
Calif.: RAND Corporation, RR-543-OSD, 2014.

experts) approaches overall, an emerging "trend in the evolution of demand forecasting is applying a combination of such techniques."[2] This combination could be approached either from the top down or the bottom up.

A top-down approach uses historic workforce data for regression analysis to "derive a relationship between the number of workers required and measures of workload."[3] The future workload projection is then generated using historical data on demand drivers. While a standard regression analysis describes demand drivers in quantitative terms, such terms may not fully reflect actual workload. Therefore, to more fully capture the total workload, qualitative methods may be more appropriate. Note that, as before, this approach assumes that external or internal factors will not lead to a significant change in workforce productivity. It also provides no insight into the reasons for a given staffing level or whether the current level is actually optimal or even adequate.

Conversely, a bottom-up approach uses "detailed estimates of the labor required for each piece of work and then aggregates the estimates to calculate the total staffing level required."[4] Those detailed labor-task estimates could be derived either quantitatively (e.g., using time studies, which are more accurate but dependent on data availability) or, more practically, qualitatively (e.g., by soliciting estimates from managers and workers, which provide insight into all aspects of a particular function but can be subject to bias). This method has the dual benefit of providing more insight into workforce size requirements and relying less on historical staffing as a guide for future estimates than the top-down approach. However, it requires more specialized and time-intensive data collection, and more importantly, it must fully capture all tasks performed by a given unit and time expenditures by every worker.[5]

---

[2] Nataraj et al., 2014, p. 69.

[3] Nataraj et al., 2014, pp. 69–70.

[4] Nataraj et al., 2014, p. 72.

[5] Nataraj et al., 2014, p. 72.

Methodology for Selecting Appropriate Workforce Demand Model for the Supply Management Units and Repairable Issue Points

To determine which workforce demand model is most suitable for an organization, the key characteristics of the functions to be modeled must be assessed. These characteristics include:

- complexity of the function
- level of task-data collection availability
- uniqueness.

When determining the complexity of the functions, it is important to determine whether all tasks are easily defined and captured. Complex functions have a plethora of tasks associated with them, and this increased complexity risks the possibility of omitting certain tasks or oversimplifying them. Less complex functions have a clear set of tasks and drivers of workload. When tasks are clearly outlined and understood, bottom-up workforce models work best. When functions are overly complex, the use of qualitative approaches are more suitable. Similar to complexity, the level of data availability and the level of data-collection automation help determine the appropriate model to be used for staffing estimates. When data availability and level of data-collection automation are high, bottom-up, task-frequency-duration models are more appropriate. Finally, the uniqueness of a function determines whether or not other functions can be used as comparisons. Uniqueness is defined as performing tasks that no other groups perform.[6]

In the case of the intermediate supply storage section, the tasks and sequencing of tasks were clearly laid out in unit SOPs. Therefore, while the storage section function is a bit complex, we felt that all tasks were systematically captured, thus allowing for the use of a bottom-up workforce model. Similarly, about half of the tasks had automatic data

---

[6] Based on James R. Broyles, Shawn McKay, Albert A. Robbert, Kristin Van Abel, Maria DeYoreo, Cedric Kenney, and Kristin J. Leuschner, *Staffing Models for Customs and Border Protection's Support Services: A Generic Methodology and Specific Applications*, Santa Monica, Calif.: RAND Corporation, RR-2553-DHS, forthcoming.

collection through STRATIS, and one of the MEFs provided detailed data on task durations that it had recently collected. This level of data availability again suggested that a bottom-up approach was the appropriate fit. Where data were missing or incomplete, SMEs were able to provide input. Finally, the storage section is unique in its functions as compared to other supply functions, so comparison models were not a good fit. It should be noted that we initially attempted to validate our bottom-up approach with a top-down approach, but the data required, most importantly historical staffing data, were unavailable, making the top-down approach impossible. Moreover, a top-down approach in this context has the additional disadvantage of assuming historical staffing levels were sufficient.

Model Mechanics

The storage section staffing model is a task-frequency-duration model that estimates the number of hours historically worked per day per storage Marine on hand.[7] For each day in the time period between January 2, 2013, and May 17, 2018, we multiplied the number of times each storage task was performed by the average time taken to complete the task. The total times for all tasks were summed and divided by the number of Marines on hand (usually a rough estimate) to estimate the average number of hours worked per Marine. An alternate approach would be to take the total person hours and divide it by the available hours per day per Marine (according to SME estimates, this would be five to six hours) to obtain an estimate of personnel required per day.

Data Collection and Inputs

The data inputs for each of the tasks in the storage model came from two primary sources and were augmented by SME input where data

---

[7] "On-hand personnel" is defined as the number of Marines at the unit available to work on storage-related tasks. It does not include Marines who are on leave, sick, or conducting other tasks such as standing watch.

were missing.[8] Table 3.1 illustrates the sequencing of tasks based on I MEF Intermediate Supply SOP and their associated frequency and duration. With input from SMEs, we identified six primary storage tasks: receiving, stowing, picking, packing, shipping, and inventory. Each of these tasks has associated subtasks, which are completed in a sequenced order. Some tasks are dependent on other primary tasks before they can be completed. For example, receiving occurs before stowing and picking is followed by packing and then shipping. The primary tasks of receiving, picking, and inventory can occur simultaneously, which is why they are broken down into sections in Table 3.1.

To determine task duration, each task was associated with an average task time. The primary source for this information was a III MEF time study conducted in 2017. Unfortunately, not all of the tasks we identified were included in that study. For those tasks that were not included we used SME input and RAND estimates based on interviews with storage experts. A key limitation of using SME input is that it is subject to bias.[9] Additionally, the times provided for these tasks were given as an average time spent per day on each type of task, rather than the average duration of the task each time it was completed.

The final input needed to complete the task-frequency-duration model was the frequency with which each task was completed. To determine the frequency each task is performed per day we used

---

[8]   Data collection across all three MEFs was highly variable and inconsistent. That led to numerous complications when creating the model. The final model was the result of several iterations and multiple data collection efforts. Recommendations for better, more systematic data collection will be presented at the end of this chapter and in the report's recommendations.

[9]   For example, Michael M. Roy, Nicholas J. S. Christenfeld, and Meghan Jones, "Actors, Observers, and the Estimation of Task Duration," *Quarterly Journal of Experimental Psychology*, Vol. 66, No. 1, 2013, pp. 121–137, found that bias in estimation was related to the duration of the task, with overestimation for short tasks and underestimation for longer tasks. Rafay A. Siddiqui, Frank May, and Ashwani Monga, "Reversals of Task Duration Estimates: Thinking How Rather than Why Shrinks Duration Estimates for Simple Tasks, but Elongates Estimates for Complex Tasks," *Journal of Experimental Social Psychology*, Vol. 50, January 2014, pp. 184–189, found that asking individuals to think about "how" versus "why" a task would be completed changed their estimates of task duration.

Table 3.1
Storage Task Sequencing and Data Sources

Function	Order	Task Name	Average Task Time (Minutes)	Standard Deviation of Task Time	Average Total Time per Day for II MEF (Minutes)	Source of Task Time	Number of Times Performed	Source of Number of Times	Section
Receiving Activities	1	Storage—Receiving	13.3	1.60		III MEF Time Study	1/Stow	I MEF SOP	Receiving
	2	Care and Storage	105		105	SME Input	1/Day	I MEF SOP	Receiving
	3	Causative Research	30		100	SME Input	5% of Stows/Day	I MEF SOP	Receiving
	4	MHE Operations	15		267	SME Input	25% of Stows/Day	RAND Estimate	Receiving
	5	Quality Control Process	3		451	SME Input	25 Items/Day	I MEF SOP	Receiving
	6	Storage Stow	5.3	2.53		III MEF Time Study	1/Stow	I MEF SOP	Carousels, Bins, Bulk
	7	MHE Operations	15		267	SME Input	25% of Stows/Day	RAND Estimate	Carousels, Bins, Bulk
	8	Quality Control Process	3		451	SME Input	5 Stows/Day	I MEF SOP	Carousels, Bins, Bulk

Table 3.1—Continued

Function	Order	Task Name	Average Task Time (Minutes)	Standard Deviation of Task Time	Average Total Time per Day for II MEF (Minutes)	Source of Task Time	Number of Times Performed	Source of Number of Times	Section
Fulfillment Activities	1	Storage Picking (by Paper) All Sections	7.3	1.30		III MEF Time Study	1/Pick	I MEF SOP	Carousels, Bins, Bulk
	2	MHE Operations	15		267	SME Input	25% of Picks	RAND Estimate	Carousels, Bins, Bulk
	3	Quality Control Process	3		451	SME Input	5 Locations/ Day	I MEF SOP	Carousels, Bins, Bulk
	4	Storage Packing	3.1	0.79		III MEF Time Study	1/Pick	I MEF SOP	Packing
	5	Quality Control Process	3		451	SME Input	25% of Picks/ Day	I MEF SOP	Packing
	6	MHE Operations	15		267	SME Input	25% of Picks	RAND Estimate	Packing
	7	Storage Shipping	14.2	1.55		III MEF Time Study	1/Pick	I MEF SOP	Shipping
	8	Quality Control Process	3		451	SME Input	40% of Picks/ Day	I MEF SOP	Shipping

Table 3.1—Continued

Function	Order	Task Name	Average Task Time (Minutes)	Standard Deviation of Task Time	Average Total Time per Day for II MEF (Minutes)	Source of Task Time	Number of Times Performed	Source of Number of Times	Section
Inventory Activities	1	Storage Inventory	8	3.48		III MEF Time Study	1/Inventory	I MEF SOP	Operations
	2	Causative Research	100		100	SME Input	1/Day	RAND Estimate	Operations
	3	Rewarehouse	186		186	SME Input	1/Day	RAND Estimate	Operations
	4	Change Location	80		80	SME Input	1/Day	RAND Estimate	Operations

SOURCE: Based on III MEF Capacity Study, I MEF SOP, and SME Input.

the I MEF SOP as our primary source. For those tasks where a frequency was not included (e.g., operation of material handling equipment [MHE]), we made an estimate based on interviews with storage experts. These assumptions are easily replaceable if better data collection occurs in the future.

Once the frequency of the task was determined, historical data sources were used to determine how often each tasked was performed each day. The primary data source used was STRATIS. This provided historical data from January 2, 2013, to May 17, 2018, for the number of stows, picks, and inventories per day. For those tasks not included we used a total time per day as related to us by SMEs. This is one of the primary limitations in the model, as the frequency and duration for these eight tasks are static.[10] It is recommended that the Marine Corps pursue better data collection for these select tasks.

The final portion of the model that required data input was the historical number of personnel on hand for the storage section. Currently, the Marine Corps lacks the ability to archive daily on-hand personnel numbers for the company and below. While this information is logged daily on the morning report on Marine Online, it is not archived. Each of the MEFs had different data collection techniques, which made it difficult to readily access this information. The most complete data source was the III MEF Capacity Study.[11] It annotated how many personnel were available by section. Using this, we calculated an average percentage of personnel on hand over the 12-week period and multiplied that by the section's T/O. To make the model more accurate, having actual data on personnel on hand is recommended.

Because many of our model inputs, particularly task times and on-hand personnel, were supplied by III MEF, we focused our efforts on modeling III MEF's workload. Since storage practices are

---

[10] These eight additional tasks identified by SMEs included quality control, sending items to DRMO, MHE operations, causative research, care in storage, location verification, change of location, and rewarehousing.

[11] This study is described by Combat Logistics Regiment 35, 3rd Marine Logistics Group, "Projecting Combat Service Support Through Capacity, Utilization, & Integration," information brief, October 27, 2017.

fairly similar across the three MEFs, we expect a similar approach to be applicable for I and II MEF if on-hand personnel data become available.

The mathematical formulas we used to calculate each task time are outlined below.

Total Stow Time per Day = (number of stows) ×
[avg. receiving time per stow + avg. stow time per stow + 0.05 × (avg. causative research time per stow) + 0.50 × (avg. MHE operations time per stow)] + 30 × (avg. quality control time per day) + avg. care and storage time per day

Total Pick Time per Day = (number of picks) ×
[avg. pick time + avg. pack time + avg. ship time + 0.50 × (avg. MHE operations time) + 0.65 × (avg. quality control time)] + 5 × (avg. quality control time)

Total Inventory Time per Day = (number of inventories) × (avg. inventory time) + (avg. causative research time) + (avg. rewarehouse time) + (avg. change location time)

Average Daily Hours per Marine = Total Stow Time per Day + Total Pick Time per Day + Total Inventory Time per Day ÷ (# of authorized enlisted personnel) × (avg. fraction on hand)

Model Results

The storage model results for III MEF are shown in Figure 3.1. Each point represents the average number of hours worked per on-hand Marine on storage tasks each day. Green squares correspond to days for which exact on-hand data was available from the III MEF Capacity Study.[12] From these on-hand data, we determined that the average number of enlisted Marines available to perform storage tasks each day was only about 42 percent of authorized personnel. This equated to 46

Figure 3.1
III Marine Expeditionary Force Hours Worked per Storage Marine per Day, January 2013–May 2018

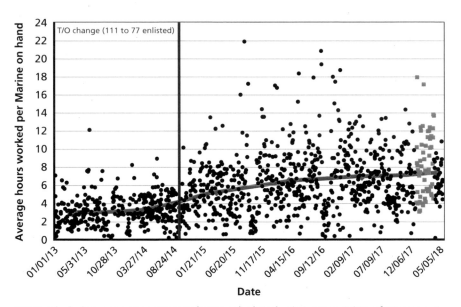

NOTE: Black dots represent average hours calculated using assumption of 42 percent personnel on hand; green dots represent accurate on-hand data provided by III MEF.

---

[12] This time study reflected only a 12-week period from January to April 2018. To show the change in workload over a period covering the full range of STRATIS data available, we calculated an average on-hand percentage that was multiplied by the T/O authorized strength to provide an estimate for personnel on hand. It was important to show this longer time period to estimate the impact the change in T/O had on hours worked.

Marines on hand before the T/O change and 32 Marines after the drop in authorized personnel. We used these numbers to calculate average hours worked per day when on-hand personnel data were not available; these results are shown as black circles. The red line in the figure represents the average number of hours worked each day over a fiscal year. The vertical blue line denotes the FY 2015 force reduction, when the number of authorized enlisted Marines in the III MEF storage section fell from 111 to 77.

The estimated average daily hours have clearly risen over the past four years, from about three hours in FY 2013 to a little over seven hours in FY 2018, with a visible jump immediately following the FY 2015 force reduction. While average hours worked might appear low at first glance, we emphasize that we do not expect warehouse Marines to spend eight hours per day on storage tasks alone. A Marine's daily workload includes tasks that are not directly related to warehouse functions, such as annual training, weapons training, enlisted PME, and MOS skills training, and therefore are not captured by our model. Our discussions with SMEs suggest that Marines would ideally spend an average of five to six hours per day on storage tasks to account for these extraneous tasks and to allow for surge capacity when demand is highest. Our model suggests that not only did the average daily hours exceed the ideal number in FY 2018, but on 36 percent of workdays, storage Marines spent more than eight hours on storage tasks alone. Statistics for FY 2018 (through May 17, 2018) are shown in Figure 3.2 and summarized in Table 3.2.

Note that the days with excessively high averages—especially those above 12–14 hours—are most likely an artifact of the model, since we did not have actual on-hand data for most of the dates in the time frame under consideration. Thus, we are assuming a fixed percentage of the personnel on the T/O are performing storage tasks each day, when in practice, fewer Marines may be released for other duties on days when the workload is heaviest. In addition, we assume that some tasks (such as causative research and quality control) are performed a fixed number of times each day or in proportion to other workload, which may not be true on days when demand is high. With better data on task frequency and duration and the actual number of personnel on

Figure 3.2
Distribution of III Marine Expeditionary Force Hours Worked per Storage Marine per Day, FY 2018

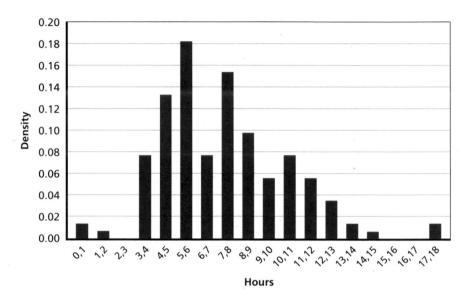

Table 3.2
Cumulative Frequency of Hours Worked per Marine per Day

Hours Worked	Percentage of Days	Cumulative Percentage of Days
< 4	9.8	9.8
< 5	13.2	23
< 6	18	41
< 7	8	49
< 8	15	64
< 10	16	80
< 12	13	93
< 15	6	99

hand each day, the fidelity of the model could be improved. To explore the effects of varying these assumptions, we performed a Monte Carlo simulation that randomly sampled task times according to a normal distribution about the mean and allowed for variation in the number of Marines on hand.[13] These results are reported in Appendix B.

Force Structure Model

The results above suggest that the III MEF storage section is almost, but not quite, right-sized. Currently, the section operates with too few Marines to meet demand while also achieving the ideal average of five to six hours worked per day on storage tasks. Fortunately, the Marine Corps has multiple options for decreasing hours worked to a more manageable level. Reducing the degree to which the section is taxed—i.e., allowing the section to operate with more than 42 percent of its authorized personnel—would decrease the number of daily hours required of each Marine on hand. Alternatively, if we assume the same percentage of Marines will continue to be pulled away to other duties, increasing the number of authorized Marines would achieve a similar result.

In either case, the model above can help us determine a more appropriate number of Marines to keep on hand. In FY 2018, the III MEF storage section had an average of 36 Marines on hand each day, and each of them worked an average of 7.3 hours on storage tasks. Assuming the same number of total daily man-hours is required to continue meeting demand, about 44 to 53 Marines are needed to achieve the target average of five to six hours per day.[14] If, on average, 42 percent of total authorized Marines are available, this would mean the number of authorized enlisted Marines in the III MEF storage section should be between 104 and 125. (Currently the section is authorized 86 enlisted Marines.) These results are summarized in Table 3.3.

---

[13] We used data collected as part of the III MEF time study as the basis for variations in task duration and Marines on hand.

[14] We compared these results to the alternative approach presented earlier, in which total task time was divided by number of hours available (five to six hours) and found similar results. According to that model, the number of Marines needed for a five- and six-hour day were 54 and 45, respectively.

Table 3.3
Target Numbers of On-Hand and Authorized Marines
(III Marine Expeditionary Force Storage Section)

Target Hours Worked	Number of Marines Needed on Hand (current average = 36)	Delta Between Marines Needed and Current Staffing	Number of Authorized Marines Needed Assuming 42% Availability
5	53	17	125
6	44	8	104

A shortcoming of the calculations above is that they assume future demand will be similar to the FY 2018 demand, which might not be the case. In order to develop a more accurate storage section model that is adjustable for changing demand, we would need to establish clear relationships between the number of hours worked per Marine, demand, and drivers of demand, such as the number of units supported and their sizes or the number of exercises planned for a given time period. Unfortunately, until accurate on-hand data become available, the relationship between hours worked and demand cannot be clearly established. Likewise, we were unable to acquire the necessary data to capture certain drivers of demand.[15]

We were, however, able to determine the number of units supported by the III MEF storage section each day from GCSS-MC requisition data by counting the number of unique unit identification codes, which we compared to the number of daily GCSS-MC records, a proxy measure of demand. There exists a positive correlation (about

---

[15] For instance, we did gather data on the frequency of exercises over calendar year 2018 for I MEF using their historical Training Exercise and Employment Plan (TEEP); however, because we lacked good data for hours worked for I MEF, it was difficult to correlate these data with SMU workload. We were unable to locate the TEEP for the other two MEFs. Additionally, we had only the duration of the exercises, not their size; the latter would have further helped us estimate demand. It is recommended that the Marine Corps locate and analyze historical TEEPs to determine if there is a correlation between exercises supported and demand, as in the approach taken for units supported. There is no central repository for historical TEEPs, so these will need to be located on unit sharedrives.

0.66) between demand and the number of units supported daily. A similar analysis for other drivers of demand would help elucidate which factors contribute most to increasing the burden on the storage section. Once on-hand data become available, these drivers could then be linked to daily hours worked per Marine in order to develop a formula that would allow the Marine Corps to adjust the size of the storage section based on the anticipated level of demand.

Additional Intermediate Supply Models

The majority of the intermediate supply workforce resides in the storage section. That is why we focused on testing our model on that section. We recommend that other sections be modeled using similar approaches. Unfortunately, we did not have access to as much detailed data for those as we did for the storage section, so we have not been able to implement these models.

Assuming that the Marine Corps can expand on existing data collection regarding the length of time needed for each task, the number of times it occurs each day, and the number of personnel on hand, we recommend a time-frequency-duration model for each of these sections. Table 3.4 shows the drivers of workload we have identified for each section. For each of these tasks, data should be collected on the average time needed to complete them, as in the III MEF Capacity Study for storage tasks. For example, in the GA section, data should be captured on the time it takes to complete keypunches for each type of transaction or to research discrepancies.

To measure the frequency of tasks, again it is important to tie each task to transactions in automated information systems, such as STRATIS, GCSS-MC, and other data sources. Capturing the variability in task frequency is important to determine capacity. Finally, we recommend that the Marine Corps archive data on the number of on-hand personnel by section, which is already being collected on a daily basis. As we discussed in relation to the storage section, accurate data on the number of personnel available to perform core tasks each day will ensure greater accuracy of capacity models.

Table 3.4
Drivers of Workload for Intermediate Supply Sections

Section	Tasks
Storage	Shipping/receiving
	Stowing
	Picking
	Verification
MAGTF Materiel Distribution Center	Receiving
	Distributing
Deployment Support Unit	Forward deployed unit support
	Build/manage Class IX blocks
General Account	Stock control
	DASF management
	Special projects
Packing	Pack
	Outbound receiving
	HAZMAT processing
	Box building
Customer Service	Reconciliation
	Walk-through
	Shipping discrepancy
	GCSS-MC support
	T&R support
Fiscal	Obligations
	Expenses
	Liquidation

SOURCE: Based on Total Force Structure Management System.

An example of the recommended approach applied to the GA section is presented in Table 3.5. Most of the task frequency data could be obtained from GCSS-MC or STRATIS, although some manual data collection would be required for a few tasks, as well as for average processing time for each task. Multiplying the frequency and duration of each task and summing across tasks will result in the total time spent by the GA section on administrative tasks each day. This total can then be divided by the personnel on hand to estimate the hours worked per person per day on GA activities.

Table 3.5
Structure of General Account Capacity Model

Task	Frequency	Duration
DASF—Lonesome demand	Number of lonesome demands	× Avg time
DASF—Aged Shipment Status (AS1)	Number of AS1s > 30 days	× Avg time
DASF—Lost shipment	Number of lost shipments	× Avg time
DASF—Follow-ups	Number of requests w/ aged est. ship date	× Avg time
DASF—Cancellations	Number of cancellations	× Avg time
DASF—Process receipts	Number of receipts	× Avg time
DASF—Modifications	Number of modifications	× Avg time
DASF—Review Materiel Obligation Validation Responses (AP1)	Number of documents w/out AP1	× Avg time
DASF—Handle Last Known Holder (LKH)	Number of document numbers	× Avg time
DASF—Proof of delivery	Number of receipts	× Avg time
DASF—Frustrated gear	Number of receipts w/out paperwork	× Avg time
DASF—Rollback	Number of rollback requests	× Avg time
Remove condition code F	Number of condition code F items	× Avg time
Stock control—Code blanks	Number of blanks	× Avg time
Stock control—Recode NIINs	Number of NIINs requiring recoding	× Avg time
Stock control—IIP assistance	Number per day	× Avg time
Special projects—MVGLs	Number of MVGLs	× Avg time

Recommended Way Forward

While modeling the storage section, we found that current data collection is inconsistent across the three MEFs. Additionally, certain tasks such as MHE operations, care and storage, causative research, and quality control lack data on individual task duration. It is recommended that more accurate and complete data be collected for capacity modeling. This includes capturing variation in tasks per day and eliminating fixed times per day. Storage sections should also consider archiving daily records of Marines on hand to improve personnel counts.

While the modeling for the supply section focused on garrison storage activities, it is recommended that staffing forecasts focus on what is required to do deployed operations. In a deployed environment, the expectation is that workload will increase (by at least a factor of two) as there is increased time required to complete tasks and an increase in volume. Finally, the storage model should be replicated for the other intermediate supply sections. A recommended application is presented in the next chapter.

Conclusion

Determining workforce demand can be performed quantitatively, qualitatively, or, as has become increasingly common, via a combination of both. To determine the most appropriate workforce demand model, one should focus on what the function of the model is. From there the function to be modeled should be assessed for its complexity, availability of automated data, and uniqueness. The storage section was assessed to be a unique, noncomplicated function with limited to good data availability. For these reasons, a bottom-up task-frequency-duration model was used to determine historical hours worked per day per Marine in the storage section. This understanding of workload will assist planners in determining the appropriate staffing level to meet demand. Our model determined that storage sections are understaffed by 8–17 personnel, assuming that the ideal work day per Marine is five to six hours on storage-specific tasks. This shortfall can be mitigated either by an increase in authorized or staffed personnel or by gaining greater efficiencies in work performance. The next chapter focuses on challenges that limit workforce efficiency and on best practices that could be shared across MEFs to increase efficiency in garrison.

CHAPTER FOUR

Challenges and Best Practices

During our site visits to the three MEFs, we gathered information about challenges that limit productivity in intermediate supply units and best practices that could be followed to improve productivity.[1] We found that many of the challenges were common across the three MEFs. In the next section, we discuss these common challenges. In the following section, we discuss best practices and unique challenges at the three MEFs. We have not attempted to quantify or verify the productivity effects of these challenges, so additional analysis may be needed to substantiate them.

Common Challenges

The issues raised by our interviewees indicate that the three MEFs are experiencing similar systemic challenges that impact their efficiency and effectiveness. These challenges can be categorized as personnel, equipment, or external factors.

Personnel Challenges

Personnel stability was mentioned by the SMEs we interviewed as a Marine Corps–wide challenge. SMEs were concerned that the current number of Marines available to perform intermediate supply functions

---

[1] In the previous chapter we focused specifically on the storage section, but here we once again examine all intermediate supply functions for challenges and best practices.

is slightly below the number required to maintain current capabilities and meet demands. Additionally, interviewees reported a mid-level supervisor shortage (in pay grades E-5 to E-7) due to reductions in force structure and assignments to other MOS-independent functions, such as recruiting duty.[2] The need to utilize personnel with the incorrect MOS or pay grade causes extra strain on the managerial section, shifting focus from process improvement to basic supervision and coaching. It also prevents managers from addressing using units' more unusual problems and exceptions to normal intermediate-level supply processes. These deficiencies, which can impact productivity and continuity, are partly offset by the addition of civilians and contractors, who provide stability to the fluctuating Marine personnel.

When a MEU or other large unit is deployed, the civilian and contractor staff maintain a regular government schedule, providing a stable workforce for the reduced activities on the base. Some of these contractors and civilians previously served in the Marine Corps and have returned as government employees, so they can provide additional expertise and guidance for newer Marines. Although it is beneficial to have some civilian and contractor personnel performing the more complicated types of workload (e.g., in the fiscal and DSU sections), it is also necessary to have Marines working in these sections, so that they will have the experience and training necessary for carrying out these tasks in the field and during deployed operations. In addition, unlike the case with military personnel, the availability of civilian and contractor personnel can be affected by government shutdowns.

In addition to the limited number of Marines available for intermediate supply work, the training required to develop the knowledge and expertise the position requires reduces the time a new Marine has available for work. Unlike the skills required for other supply jobs, the skills required for intermediate supply functions are not extensively taught in the schoolhouse and require on-the-job training. SMEs inter-

---

[2] Based on an examination of SMU T/O reductions between FY 2013 and FY 2018, the number of E5–E7 positions was not cut as deeply as other pay grades, so the issue may be related to fill rates, or taking on additional management responsibilities due to larger percentage reductions in the number of E8 and officer positions in most of the MEFs.

viewed at all three MEFs said that there is a need for more specialized training for intermediate supply units so that when new Marines arrive, they have a higher level of autonomy. Interviewees reported that a significant proportion of a Marine's three-year rotation in an intermediate supply unit is spent on learning new processes and training new Marines who are unfamiliar with the processes. While we did not assess the exact time devoted each day to coaching and training new Marines, we witnessed the practice of pairing experienced Marines with new ones and many coaching interactions taking place. The use of trained civilians and contractors can provide more stability in a section's experience level.

Last, delays in acquiring permission to access the nonclassified internet protocol router network (NIPRNet) can limit a Marine's ability to work in certain positions in either the RIP or the SMU. Since STRATIS is located on the NIPRNet, Marines need to submit access requests before they can work on tasks that utilize the warehouse system. Additionally, Marines must have a secret clearance to access the supply system's inventory. Delays in receiving access forces managers to assign Marines to positions in the warehouse where they either do not have to use STRATIS or can be partnered with someone who already has access. This challenge has a high potential to limit the intermediate ground supply system's capacity, as STRATIS is heavily integrated into the SMU and RIP process.

Equipment Challenges

Both the SMU and the RIP require MHE for transporting and stowing items in warehouses. However, the equipment we observed on our visits is typically fully depreciated, lacking, or insufficient, resulting in the need for high levels of manual labor and additional time to carry out basic tasks. SMEs we interviewed also mentioned these challenges, but often remarked that "Marines make do." Some of the equipment is shared across multiple units and belongs to the base rather than the intermediate supply section. A high percentage of the needed equipment would be considered essential in a standard industrial setting. One source of this problem may be that the SMU and RIP are hybrids of garrison operations and fully deployable units, so the unit's table of equipment may only include equipment for deployed operations. A separate list of equipment for garrison operations may be needed to ensure the availability of proper MHE, which becomes even more crucial if most of the MEF deploys and the remaining personnel need to play a support role with reduced staff. While we could not conduct a detailed time study to measure the productivity impacts of inadequate MHE, the time we spent observing processes and shadowing Marines as they performed their tasks indicated that labor hours per task could be reduced significantly with modern warehouse equipment. Better equipment that allows Marines to perform tasks more quickly could reduce the overall labor requirement and help in time of surges.

Additionally, Marines require PPE to ensure workplace safety in the intermediate ground supply warehouses. This protective gear includes safety boots, as well as eye and ear protection. However, many warehousemen whom we interviewed said they do not have access to appropriate PPE and perform tasks with insufficient protection, which could place them at significant risk of possible long-term health impacts. Although there is operational funding specifically designated for safety equipment, Marines we spoke with indicated these funds are insufficient to cover purchasing safety boots and other PPE. We did not observe Marines using hearing protection during our visits, nor did we see any lockers or storage areas for PPE distribution. When we asked individual Marines about them, none could point to where they were located.

The insufficient MHE available to warehouse Marines can lead to a greater risk of long-term health issues because of poor ergonomics. At multiple sites, we witnessed Marines climbing mobile stairs to reach upper storage racks and retrieve heavy or bulky items that required both hands to carry down the stairs. While this may sound benign, it is not accepted in normal warehouse operations, because it puts the workers and the equipment at risk. Studies of commercial workplace injuries indicate that warehouse workers suffer a high rate of back strain and sprains, resulting in significant worker compensation costs.[3] Although Marines have a limited rotation in intermediate

---

[3] A 2001 estimate valued worker compensation due to overexertion injuries at $12.5 billion. "Ergonomics: The Backbone of a More Productive Warehouse," Ergonomics Supplement. *Modern Materials Handling*, April 2004, p. S4.

supply units, routine processes carried out for a significant period of time with poor posture can negatively impact a Marine's performance and overall work capacity.

Marines we interviewed also mentioned information technology (IT) system challenges that hampered their productivity or increased customer wait time. For example, GCSS-MC is sometimes slow and unresponsive due to insufficient server capacity for the number of users or internet connectivity issues. System issues were reported as a key problem faced daily in both SMUs and RIPs, as well as in using units. One using unit in Okinawa reported having a Marine working full time simply trying to get better visibility of items that are either exceptions in GCSS-MC or absent from the system. An analysis should be conducted to determine the root cause of software issues and then prioritize remedies that would have the greatest impact on improving the effectiveness of Marines using the system.

External Challenges

In discussions with members of both the SMU and the RIP, we repeatedly heard funding consistency mentioned as a challenge facing the intermediate supply system. The seasonality of the budget cycle prevents management from effectively forecasting and establishing a spending rate that provides the greatest support possible to using units. A climate of continuing congressional resolutions, funding instability and spending restrictions can cause system shocks in the supply chain and lead to inventory shortfalls and backorders.

Additionally, fundamental changes to the intermediate supply structure can affect the ability of the RIP and the SMU to provide support to using units. Recently, the Defense Logistics Agency (DLA) has initiated three different proofs of principle (PoPs) at each of the MEFs to assess the feasibility and practicality of reducing the workload for the SMU by shifting responsibility for some of the NIINs currently stocked by the intermediate supply section to DLA. While these initiatives may reduce the overall costs for the Marine Corps, they may also result in increased lead time for units during critical moments of high demand, such as deployments, if reductions in inventory levels are implemented based on peacetime demand patterns.

Best Practices and Unique Challenges

During site visits, we discovered that while the SMUs and RIPs at each MEF have similar structures and procedures, there are slight variations in processes from unit to unit. Each of the intermediate supply sections at the three different locations have implemented some best practices that would be valuable to expand to the other units. We also document some of the unique challenges facing each of the units that we identified.

I Marine Expeditionary Force
Supply Management Unit

I MEF is the only MEF that is International Organization for Standardization (ISO) certified. To remain compliant, the SMU emphasizes quality management over its processes. It must maintain detailed and up-to-date versions of its SOPs, which it audits every six months and internally checks in a formal monthly review. The primary benefit of ISO certification, according to stakeholders, is that the rigidity of the ISO program and the documentation required to make process changes force the organization to maintain operational continuity even when managers are replaced. Marines indicated that at other units where ISO certification is not conducted, a change of management often brought a lot of new ideas on how to do things, which after a few rotations proved chaotic.

In addition to monitoring its processes to maintain its ISO certification, the I MEF SMU also utilizes Lean Six Sigma (LSS) methods to increase efficiency. The SMU has a distinct section dedicated to process improvement; however, every Marine is at least a yellow belt and is encouraged to recommend areas for improvement.[4] The SMU is

---

[4] An LSS yellow belt requires training in the basic aspects of the LSS methodology, including the phases of Define, Measure, and Control and qualifies the individual to lead limited improvement projects or serve as a team member on more complex improvement projects led by others with higher-level LSS qualifications (green belt or black belt). A green belt requires a more thorough understanding of the core to advanced elements of the LSS methodology, including the phases of Define, Measure, Analyze, Improve, and Control. See, for example, International Association for

training other units within I MEF to emphasize process efficiency by providing internal LSS green belt courses. The adoption of LSS provides a source of pride to the Marines in the SMU and demonstrates a very successful program that could be replicated in the other MEFs.

However, the SMU at I MEF does face some challenges. The SMU is split across several warehouses and buildings in the middle of the base. Almost all sections are situated in separate warehouses that are lined up along former rail lines. As these legacy warehouses were created for rail storage, they were not specifically designed for the current intermediate supply process. Whereas modern warehouses are deeper and resemble a broad boxy rectangle, the current setup affects the ability to establish efficient storage footprints and lengthens the average distance that must be traversed to retrieve items. In combination with limited MHE, this configuration leads to inefficient use of time spent walking between warehouses. While changing the physical layout of warehouse would be a major endeavor and could not easily be addressed in the short term, it is an important factor to consider when looking at task time and manpower requirements. Increasing the distance that personnel need to walk or transport goods adds to the labor hours required for the task. Measuring the exact productivity impact of this factor was beyond the scope and resources of this study. However, it is something to bear in mind when benchmarking the man-hours needed to complete similar tasks in more optimal settings.

Moreover, most of the warehouse equipment is outdated. In the primary SMU warehouse, the conveyor system that is used to send small consumables to and from the bins section is frequently inoperable. The contractor is in New York, and maintenance is costly and difficult to schedule. When the system is down, the Marines' yield and their overall capacity is greatly diminished as they must rely on more manual solutions.

In contrast to typical SMU processes at the other MEFs, I MEF opens some packaged bulk items to confirm the listed quantity and to package each item separately. For example, a Marine might open a

---

Six Sigma Certification, "Yellow Belt Certification" and "Green Belt Certification," 2018a and 2018b.

package of 100 rifle-cleaning brushes and place each one in a small plastic bag prior to sending them to the storage subsection. This requires a significant amount of Marine manpower compared to stowing the entire package and picking the requested number of items when an order is received.

Repairable Issue Point

I MEF's RIP is part of the Repairable Maintenance Company (RMC). The RMC is housed in a compound that includes the RIP and a few maintenance warehouses. The RIP, including its administrative offices, is housed entirely within a single warehouse adjacent to the ocean. The sea climate and salt in the air result in increased rust and other environmental damage to the equipment. Special care is needed to prevent corrosion, including increased packing standards. This situation has led to an initiative to provide additional capability to preserve and package (P2) items. Although a P3 satellite is located nearby, the RIP has its own P2 standards to protect items against corrosion and storage degradation when they return from repair. Prior to this initiative many RIP items would corrode on the shelf and could not be issued. Instead they had to be returned for a second repair. The P2 initiative has reduced the number of double repairs required due to corrosion.

One challenge for the RIP is that it has limited control of its budget execution. The Marine Logistics Group (MLG) and MEF G-8 (Comptroller) give the I MEF RIP month-to-month spending targets, requiring it to balance its own functions with the needs of the larger MEF. The MEF tries to maintain a steady spending rate, and, as the RIP often makes substantial purchases, the MEF utilizes it as a buffer to quickly stabilize deviations in the MEF's spending rate. As a result, this RIP lacks the ability to efficiently manage operational needs, which can result in an inefficient financial bullwhip effect.[5]

---

[5]   GCSS-MC data indicate that over the period from FY 2015 to FY 2017, the I MEF RIP purchased an average of 200 SECREP repairs per month from third-party logistics contractors, with a standard deviation of 97. II MEF RIP, which has greater control over its budget, had a lower standard deviation of 75, relative to an average of 176 contractor repairs per month.

II Marine Expeditionary Force
Supply Management Unit

II MEF has greater flexibility than does I MEF in how it spends its funding. Since they have access to their budget for the whole year, they can focus on the larger operational needs of using units, instead of solely addressing short-term issues. This best practice provides better financial management for both the SMU and the RIP. We recommend that the other MEFs implement similar financial practices.

The II MEF SMU is consolidated in one warehouse, with the exception of Initial Issue Provisioning (IIP) and HAZMAT items. A large conveyer system runs through the entirety of that warehouse; however, the work flow in the building was reversed a few years ago, rendering the conveyor system useless. The SMU also has a poor layout for sorting and packaging as the packing section lays out the packing boxes on the floor. Marines are forced to bend over for long periods of time and work on their knees in positions that could result in long-term disability (see Figure 4.1). This setup has been their de facto method of sorting outbound shipments. A modest investment in a proper sorting station would improve their efficiency and avoid future medical and disability costs.

Repairable Issue Point

To ensure that the equipment it receives is properly configured, II MEF RIP borrows specialists from the intermediate maintenance activities. This practice minimizes induction and determination errors and ensures that items are properly configured to be sent for repair. The other MEFs rely on visual configuration guides, which could cause longer repair cycle times if RIP personnel accept items with incorrect configurations, requiring multiple interactions with using units to try to obtain missing parts.

The RIP is in a maintenance compound that is relatively far away from the SMU. Its headquarters and customer service area are in a facility that is disconnected from the warehouse component. As the II MEF RIP is located in repurposed garages, its layout is not conducive to warehouse operations, nor is it optimized for them. Moreover, some functions occur in areas either fully or partially open to the environment, where extreme temperatures have a visible impact on personnel.

Figure 4.1
Sorting Station for the II Marine Expeditionary Force Supply Management Unit

SOURCE: RAND Corporation photo taken during a site visit.

III Marine Expeditionary Force
Supply Management Unit

The III MEF Supply Battalion has created a simple Excel model to forecast demand across the different supply sections and to plan based on current and projected capacity. This model records the different task types Marines would have to carry out as well as the normal time it would take to carry out each task. This model provides leadership with greater insight into capacity and readiness.

The SMU is located in a purpose-built warehouse that it shares with the IIP. The warehouse was specifically designed for its function, which ensures proper warehouse space and allows for increased efficiency in garrison. Additionally, as the SMU to the smallest MEF, it is able to pull two workloads a day when the initial workload pull is manageable. This arrangement equalizes warehouse workloads and smooths the demand signal from the using units. One challenge is that MHE is shared between the SMU and IIP, which requires additional planning to ensure that the necessary assignments can be carried out.

The overseas location of III MEF creates specific challenges for the Marines working at the SMU. Most Marines have a two- or three-year rotation in Okinawa, which results in a high turnover of personnel and limits the amount of personnel with the experience needed to operate the SMU at full capacity. Additionally, the Japanese government imposes specific travel restrictions on the island regarding the size of the trucks Marines can use, their speed, and the hours they can travel. For instance, USMC trucks cannot travel between the different camps from 0600 to 0900 or after 1100. The SMU relies on a commercial "white gear" contract using a local Japanese company with local Okinawa drivers. The service is shared across the island for all of III MEF. Without organic support, the III MEF SMU sometimes has to delay shipments if there is a higher priority need for the trucks elsewhere, which can limit distributions to using units.

Repairable Issue Point

The main RIP is in a warehouse near the SMU on Camp Kinser, while a small electronics-focused RIP is located at the Electronics Maintenance Company (ELMACO). This second location increases the interactions between the RIP and IMA, while also providing more drop-off choices for the using units. Camp Hansen also has a satellite RIP with one warehouseman and one administrative Marine. Since III MEF is scattered across multiple camps on Okinawa, the northern RIP satellite reduces the distance units must travel to replace their Code F SECREPs, but even with multiple drop-off locations, no RIP functions are conducted at some camps. III MEF is evaluating the possibility of expanding the number of satellite RIPs to all camps on the island to improve customer service.

Another III MEF–specific challenge is that the current 3PL provider sends Code F SECREP engines for repairs to the continental United States (CONUS) using military air cargo (with transportation cost paid by the Department of Defense [DoD]). The reason for doing so is that no vendors on the island meet repair standards and, the

IMA does not have the capability or capacity to perform the repairs. As a result, additional aircraft space, fuel, and money are consumed for repairs outside of the Marine Corps' maintenance structure and the repair lead time is further delayed.

Conclusion

Marines working in intermediate-level ground supply processes carry out the same tasks across the three MEFs. However, service-wide and MEF-specific challenges impact how each RIP and SMU conducts its processes. These differences provide the MEFs greater flexibility to become as effective and efficient in their supply processes as possible in their environments. However, opportunities for improvement remain, and many industry best practices can be applied to the warehouse operations at both the RIPs and SMUs. Changes, such as providing proper MHE and access to PPE, investing in modest facility upgrades, and minor adjustments in processes can increase garrison efficiency without degrading overall mission capability. We discuss our recommendations for process improvements in Chapter Five.

Findings and Recommendations

During our site visits to the three MEFs, we identified challenges facing intermediate supply units that prevented them from operating at peak efficiency. Most notably, while the units for the most part conduct similar tasks across the MEFs, each MEF individually establishes its own processes and procedures. We identified areas where improvements can be made across the MEFs to increase effectiveness.

Using available data, we were able to develop a capacity model that presents a fairly accurate picture of the SMU storage section, but data limitations prevented us from developing similar models for the other sections. Additional data collection efforts, such as archiving daily records of the number of Marines on hand in each section and measuring the average duration of the tasks performed by each section, would enable the Marine Corps to develop more accurate capacity models and better demonstrate the effects of manpower reductions in intermediate supply functions.

This chapter presents our research findings and recommendations.

Findings

This section summarizes the major findings of our research. The first group of findings is based on our modeling effort.

- **Intermediate supply storage sections are understaffed to meet current demand.**

Based on our modeling, we estimate that storage Marines in III MEF currently work an average of over seven hours a day on storage-related tasks, which exceeds the recommended six-hour work day, accounting for other training and military responsibilities. About 36 percent of the time, storage Marines work over eight-hour days to meet demand. Within III MEF, the storage section is understaffed by 8–17 personnel (22–47 percent on average).[1] If demand continues to increase, it will put further stress on intermediate supply units.[2]

- **Data collection is inconsistent across the three MEFs, and there are significant deficiencies in data required to forecast appropriate staffing requirements.**

All three MEFs collect various types of data about their processes, but they are not necessarily the best types of data for capacity and workload modeling. Of the three MEFs, the III MEF data collection effort is the most comprehensive and sophisticated. The primary deficiency in data needed for capacity modeling is daily counts of on-hand personnel by section. Other deficiencies include data on the average duration of supply tasks in sections other than the storage section and the average time to accomplish storage tasks at I and II MEF.

The second group of findings is based on our observations and SME interviews during site visits and may require additional analysis to confirm.

- **Personnel instability and lack of E-5 and E-6 personnel significantly decrease unit efficiency.**

---

[1] We ran similar modeling of the I MEF and II MEF storage sections, which resulted in similar results. However, due to the fact that we did not have time data for I and II MEF, we cannot make a similar assessment of the level of understaffing at those two units. There are some unique aspects of III MEF storage operations, such as the design of its warehouse and its ability to pull two workloads a day, which could have an impact on their time data to accomplish certain tasks.

[2] For example, at the 80th percentile of workload, Marines work an estimated 10 hours per day. To reduce the workday to 6 hours, 60 personnel would be needed— i.e., an increase of 24.

SMEs reported a shortage of mid-level supervisors (primarily E-5 and E-6 billets) at intermediate supply units. This shortage, combined with frequent turnover in personnel, reportedly results in billets being filled by personnel in the wrong MOS or pay grade. This in turn places additional strain on managerial sections to focus on training personnel and sometimes performing supply processes themselves. Supervisors may thus be prevented from fulfilling their primary duties and limited in their ability to address using units' more unusual problems, as well as exceptions to normal intermediate supply processes.

- **Significant equipment deficiencies impede the ability of Marines to perform their primary tasks.**

We observed that the MHE utilized by the SMUs and RIPs across all three MEFs is outdated, improperly maintained, and insufficient in quantity. The lack of appropriate equipment most likely results in high levels of manual labor and additional time required to complete basic tasks relative to commercial warehouses. In addition, Marines did not appear to have sufficient PPE, such as safety boots and eye and ear protection, which can lead to long-term health impacts. SMEs reported that poorly functioning IT systems also resulted in additional requirements for manual labor and increased the time required to complete basic tasks.

- **Inconsistent funding allocations cause inefficiencies.**

At some of the MEFs, SMEs reported that intermediate supply organizations are required to adjust their spending patterns to help the MEF achieve uniform budget execution over the fiscal year. This practice prevents management from effectively forecasting and establishing a funding strategy that provides the greatest support to using units. In a climate of continuing congressional resolutions, funding instability and restrictions can cause systemic shocks in the supply chain because inventory replenishment is curtailed when funding is short or accelerated to make up for underspending in other parts of the MEF. When intermediate supply organizations are allowed to control their own spending, they are better able to support using units by replenishing inventory as needed.

Recommendations

We organize our recommendations into three groups, based on whether they are related to staffing intermediate supply units, improving staffing modeling, or improving business processes in intermediate supply units.

Right-Sizing Intermediate Supply Units

- **Increase T/O or staffing goals for 3051s**

According to the model results for the III MEF storage section, in order to reduce daily work hours in the storage section to six, the number of enlisted personnel on hand would need to increase by 8–17. This can be achieved by either an increase in the number of positions on the T/O or a higher staffing goal for storage MOSs. Additional efficiencies can be gained in garrison from process efficiencies, but our primary recommendation is to increase staffing to account for deployed operations where garrison efficiencies will not have impact.

Improving Staffing Modeling

- **Archive on-hand personnel data**

On-hand personnel data by section is essential for accurate modeling of current demand and forecasting staffing needs. Currently there is no automated way to archive these data, although morning reports are recorded in Marine Online portals on a daily basis. We recommend that the Marine Corps develop a process to archive daily personnel counts at the section level.

- **Collect data on task frequency and duration for all SMU and RIP sections to improve capacity models**

A data collection effort similar to the III MEF Capacity Study is needed for each supply section and each MEF. In many cases, automated information systems can be used to measure or infer task frequency. To develop task-frequency-duration models similar to the one we created for the storage section, a comprehensive list of tasks should be created and linked to types of transactions in GCSS-MC or STRA-TIS if possible. Data on the average time to complete each task could be collected over a fixed time period. While ideally these data should be collected at all three MEFs, one of the MEFs could be used as a test case to avoid overburdening limited resources. These data collection efforts would also help managers better understand supply capacity and communicate staffing needs to other organizations.

- **Forecast staffing requirements for deployed environments**

To properly forecast staffing requirements in deployed environments, further data collection and analysis of deployed supply support for MEUs, SPMAGTFs, and other deployed organizations should be conducted. Additional data are needed on the number of intermediate supply personnel assigned to deployed units, the types and frequencies of the tasks they perform (which could be based to some extent on GCSS-MC transactions), and changes in task duration in a deployed environment.

Improving Business Processes

- **Provide basic MHE found in all U.S. commercial warehouses**

The lack of basic MHE has a tremendous impact on the velocity of operations and the time required for Marines to perform tasks safely and efficiently. There can be negative consequences in terms of lost time or workplace injuries from operating manual equipment and walking long distances in the warehouse. Neither is an efficient use of their time or of tax dollars. In addition, providing better MHE could reduce the number of Marines needed to perform daily workload.

Based on our observations, we would suggest that each MEF should invest in motorized ride-on electric pallet jacks. The ride-on jacks are less expensive than forklifts and the norm for carrying loads within a warehouse and across warehouses for short distances. They are compact and safe while providing higher velocity than walking. An electric pallet jack would roughly triple the number of movements a single Marine could accomplish with the current manual pallet jacks.

Electric forklifts designed to operate inside warehouses are also needed. The types of forklifts being used by SMUs and RIPs vary greatly across MEFs, and in many instances, they are using diesel-fueled forklifts inside warehouses, which generate excessive noise levels for workers, as well as toxic fumes. This sort of arrangement would not be allowed in a civilian warehouse. While electric forklifts are slightly more expensive than pallet jacks, providing two per SMU and per RIP would alleviate many problems. This is not about bringing Marines to a state-of-the-art future, but bringing them in line with basic equipment found in any U.S. commercial warehouse. While the bases are currently responsible for purchasing these types of equipment, it might be time for the table of equipment to be updated to reflect the basic equipment needed by Marines to do garrison work.

- **Ensure that Marines have the correct PPE to perform their work**

The Marines working in logistics MOSs deserve to have proper equipment to do their job. The lack of basic PPE and workplace safety during our visit was apparent. A sufficient budget is needed to provide proper safety boots and shoes, eye protection, high-visibility vests, and ear protection. While there is a budget currently earmarked for these items, none was available to the Marines in the warehouses we visited. Supervisors we interviewed said that the budget is insufficient. Given the small size of the workforce, workplace safety needs should be a higher priority to avoid any lost days caused by accidents. At current staffing levels, the SMUs and RIPs cannot afford losing days to preventable workplace safety issues. This recommendation would have a less direct effect on productivity than improving MHE, but could have an indirect effect if less work time is lost due to workplace injuries.

- **Allow the SMUs and RIPs at all MEFs to control their spending rate**

As discussed in Chapter Four on the site visit findings, one of the problems mentioned by SMEs at I and III MEF is the lack of full control of their budget. Most organizations like to see a relatively steady spending rate throughout the year. While the desire for predictable and stable spending is understandable, strictly adhering to it in order to have nice graphs is detrimental operationally. Adjusting spending to meet arbitrary goals instead of replenishing inventory when needed can induce a bullwhip effect, as described by Forrester in 1961.[3] While Forrester attributed the amplification of the distortion of the demand signal to poor visibility, in this case the fluctuation of the demand signal is due to the lack of funding, which creates a self-inflicted financial bullwhip.

Hence, we recommend that the RIP and SMU be free to adjust their spending rate during the fiscal year in response to customer demand, while staying within total budgetary bounds. The demand signal for repairs is not steady throughout the year due to a number of causes, including deployments, exercises, and other fluctuations in unit activities during the fiscal year. Replenishment of inventories in response to this demand should not be tied to an artificial, self-imposed constraint. This recommendation could reduce administrative workload for personnel involved with financial management, customer service, and monitoring backorders.

- **Optimize item locations at both RIPs and SMUs to reduce picking time**

We observed that item locations have mainly been established organically over the years based on space available at the time when items were received. A significant reduction in transaction times at RIPs and picking time at SMUs could be achieved by rearranging items so that the fast-moving items are as close as possible to the service counters at the RIPs and to the shipping and receiving areas at the SMUs. For example, Wulfraat reports that travel time can account for 50 percent or more of order picking hours. Therefore, he recommends that the fastest-moving items should be located at ground level and should be concentrated in "hot zones" to reduce picking time.[4] This

---

[3] Jay W. Forrester, *Industrial Dynamics,* Cambridge, Mass.: MIT Press, 1961.

[4] Marc Wulfraat, "5 Ways to Improve Order Picking Productivity," *Warehouse/DC News,* May 15, 2013.

recommendation should improve productivity and reduce the number of Marines needed to perform intermediate supply tasks in garrison.

- **Foster an LSS program for continuous improvement**

The LSS program at I MEF is very successful and in high demand across Camp Pendleton. The SMU Marines taking part in this initiative have earned a reputation for problem-solving and efficiency improvement across other organizations. It serves not only as a great motivator for the force but also as a conduit to foster continuous improvement. We estimate that it would require leadership support to stand up similar initiatives across the other MEFs. However, it would likely increase workplace pride and improve productivity over the long term.

General Intermediate Ground Supply Process

This appendix describes the general intermediate-level ground supply processes at the SMUs and RIPs of the three MEFs.

Supply Management Unit

Order Fulfillment

While using units receive consumables through the SMU either in garrison or in the field, the processes and priorities differ slightly (see Figure A.1). A using unit in garrison submits requests for consumables through GCSS-MC by NIIN. The request is transferred to STRATIS, the warehouse management system used by the Marine Corps. If the NIIN is shown in STRATIS and is available locally in the warehouse, the request will be included in the next picking and packing workload; otherwise, the request will be passed over to the nearest DLA warehouse and flagged in the system for further inspection by the GA section. If GA identifies the request as a necessary purchase, it will first confirm that the purchase fits within the SMU's budget before ordering the item.

STRATIS releases a daily workload for the items that are to be picked and packed and sent out for that day. It includes the NIINs and the quantity of items picked as well as the warehouse location where it can be found. With the exception of III MEF, which pulls a workload in the morning and in the afternoon, each MEF receives one daily workload. Warehousemen in the storage section pull consumables listed in the workload from the storage area, confirm the NIIN

Figure A.1
General Supply Management Unit Process

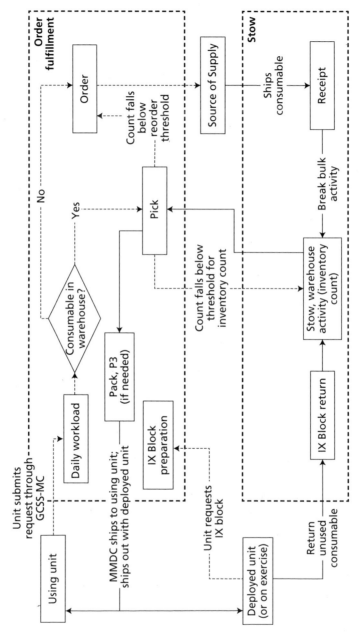

SOURCE: Generated from RAND site visits, interviews, and review of unit process maps.

and quantity listed on the workload, record the pull, and set the items aside to be transferred to the packing area. Multiple checks are taken throughout the process to ensure quality control. If a particular item requires additional packaging, it will be sent to the P3 section. If a NIIN is listed in the workload but cannot be found at the specified warehouse location, the warehousemen will perform a search in the surrounding area before reporting an error.

SMU warehouses are separated by the size of the NIIN into bins (small consumables), G-Lot (larger consumables), and bulk (bulk consumables). The bins warehouse contains electronic carousels that store small consumable items. Marines use stationary lifts to control the carousels and access the higher levels. Teams of two or three Marines operate the lift and carry out the picking and packing for a single carousel. The bins equipment across the MEFs is supplied by a contractor located in New York.

During the picking process, the packing section prepares by printing manifests it receives from STRATIS and sets up packing boxes for each customer. When the picked items are delivered to the packing station, the NIINs and their quantity are once again confirmed against the manifest before being packaged. Any picking errors are fixed at this time. The packing section then relocates packed boxes to a staging area. These are then delivered to the MMDC for distribution to the requesting units, usually the following day. Although units are allowed to do walk-throughs and pick up a consumable in person during the work day, after-hours walk-throughs are restricted to priority O-2 and higher.

IX Block

Units that are preparing to deploy can request that a Class IX Block of repair consumables be prepared to serve as a stand-in SMU in the field. The IX Block components belong to the DSU section of the SMU until the using unit uses them. Although there are standard lists based on common NIINs needed for specific environments, the unit's supply officer works with the DSU to make unit-specific modifications. Once a list is prepared, the SMU checks the warehouse to confirm that the requested NIINs are available. The picking and packing process is the same as mentioned above. Requests from deployed units are also handled by the DSU section.

Stowage

In addition to fulfilling orders, the SMU also orders new consumables, receives them, and stows them to maintain its inventories. These tasks fall under the unit's larger warehouse management function.

STRATIS has built-in thresholds for reordering and for conducting inventory counts for NIINs already in the system. While warehouse management can manually adjust these order quantity thresholds for specific NIINs at any time depending on the item's turnover rate, a standard mandatory inventory count is automatically conducted when the recorded inventory falls below ten. This is separate from the required spot checks provided by STRATIS as part of the daily workload. The warehouse CWO monitors the demand of NIINs to ensure that they are sufficiently stocked at an economical reordering quantity.

When ordered NIINs arrive at the SMU, the receiving section receives and receipts for the package by matching the manifest with its DD Form 1348 (DoD Single Line-Item Requisition Document) to confirm it. It then processes the shipment by counting the NIINs and their quantity. The NIINs are separated based on their warehouse location and picked up by warehousemen to be stowed. Items that are either not identifiable or do not match the manifest are set aside to be further investigated.

The storage section stows receipted items after the picking process is finished for the day. Before the item is stored, another quality check is performed. STRATIS indicates where the item is currently stored based on the NIIN; if this is a NIIN that does not currently have a storage location, or if the additional items do not fit in the current location, STRATIS designates a new spot. Warehousemen try to consolidate NIINs into a single location whenever possible for space efficiency.

IX Block

When a unit returns from deployment, it brings back unused consumables from its IX block to the SMU. The SMU then receipts and stows the consumables back into the warehouse.

Reconciliations and Training

In addition to order fulfillment and stowage, the SMU also performs monthly reconciliations of its using units' materiel property. The customer service section schedules appointments with a unit's supply section to confirm their records. Additionally, as the local SMEs, customer service Marines also visit units to provide supply support training.

Repairable Issue Point

Figure A.2 shows the general RIP process in place at the three MEFs. Like the SMU, the RIP interacts with a unit differently if it is garrison or it is either deployed or in the process of deployment. At the garrison RIP, the using unit delivers a configured nonfunctional (Code F) SECREP that it wants to be replaced with a functional (Code A) SECREP of the same NIIN. If the item is a controlled cryptographic item (CCI), there are additional steps to secure it. High-value and pilferable SECREPs are stored in a vault with additional security. The RIP confirms that the SECREP is properly configured with the necessary parts, and the equipment is receipted and loaded into the automated system. If there is a Code A SECREP available in the inventory of the same NIIN, the RIP will provide the using unit with the new item. SECREPs under a different NIIN or valid substitutes require further approval from the using unit's supply officer. If there is no Code A SECREP or substitute available, the item will be placed on backorder. The RIP can provide units with replacement equipment without simultaneously receiving a Code F SECREP, provided the unit returns the Code F SECREP within an allotted time.[1] The using unit inspects the Code A SECREP before receipting for it; if the unit does not accept that SECREP, the RIP issues a new item and the rejected SECREP is classified as a Code F.

Following the RIP's receipt and induction of the Code F SECREP, the administrative section of the RIP determines where it should be

---

[1] One example would be a deadlined vehicle that would require a working engine before the other engine could be returned.

Figure A.2
General Repairable Issue Point Process

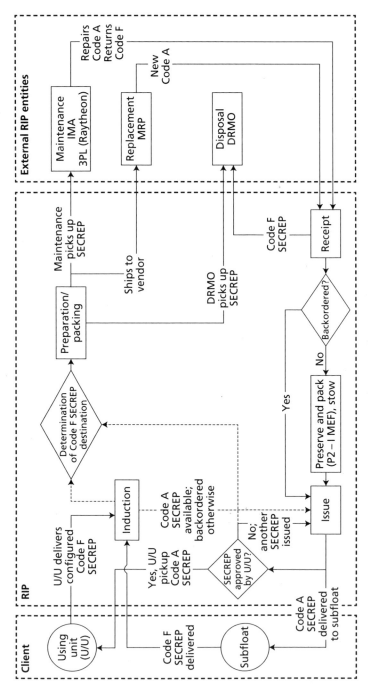

SOURCE: Generated from RAND site visits, interviews, and review of unit process maps.

sent. The options vary based on the status of the item and the RIP's current funding level:

1. IMA: The Marine Corps internal maintenance companies have the capability to repair some Code F items and restore them to Code A. They return any Code F SECREPs they cannot fix to the RIP.

2. 3PL: A contractor either repairs or overhauls Code F items and returns Code A items. At the time of our visits, this contract was funded through Overseas Contingency Operations (OCO) funding.

3. Materiel Returns Program: This program allows the RIP to send certain Code F SECREPs to maintenance depots or vendors for new Code A SECREPs.

4. DRMO: For Code F materiel that is either obsolete, beyond the point of repair, or no longer carried by the Marine Corps, the item is sent for disposal.

In cases 1 and 2, items repaired by the IMA or 3PL are returned to the RIP in Code A condition, receipted, and stowed in the warehouse. If the item cannot be repaired by the IMA or 3PL, it is returned to the RIP in Code F condition and sent to DRMO for disposal. If a Code A item is backordered, it will be issued to a using unit based on priority. In most instances, the using unit needs to return to the RIP to receive the new item.

For deployed units, a subfloat serves as the RIP in the field. When a subfloat is created, the requested SECREPs are pulled from the warehouse and sent with the unit; however, exchanges are carried out using a similar process in the field as in the garrison.

Model Evolution and Validation

Preliminary Model Results

The III MEF storage section model presented in Chapter Three is the final version of a model that was refined over time with the help of storage SMEs. To understand the degree to which some of the auxiliary tasks included in the model (Table 3.1) contribute to workload, it helps to consider a previous version of the model that does not incorporate such tasks. In this section, therefore, we present results from an earlier iteration of the model wherein we treated storage, receiving, stowing, inventory, picking, packing, and shipping (here called "primary" tasks) as the exclusive drivers of workload, ignoring "secondary" tasks such as care and storage, causative research, MHE operations, quality control, rewarehousing, and changing location.

We originally neglected the secondary tasks because we assumed time spent on them was negligible compared to the primary tasks. Data on primary tasks were also more readily available. For example, the number of times each of the primary tasks occurred daily was obtainable directly from historical STRATIS data, which indicates the number of stows, picks, and inventories performed. Primary task times were also provided in the III MEF time study. In contrast, the Marine Corps currently keeps very little quantitative data on the secondary tasks listed above. As a result, conversations with SMEs were crucial for helping us understand how many times the secondary tasks are performed daily as well as how long they take.

The results of the preliminary model are provided in Figure B.1, which shows the average number of hours worked per Marine on hand per day on primary storage tasks over the same time period as Figure 3.1. Recall that in the final version of the model in Chapter Two, the average number of hours worked per day per storage Marine in FY 2018 was approximately 7.3, after we added information from SMEs on the average daily hours needed to complete secondary tasks. Figure B.1 indicates a much lower daily average for FY 2018—5.1 hours— suggesting secondary tasks contribute substantially to daily workload. Indeed, the results help explain why relying only on the supply data that are currently available can be misleading; without secondary tasks, it would appear that the section is already operating, on average, within the ideal five- to six-hour range. The results also reaffirm the notion that improving data collection is crucial for right-sizing intermediate supply sections, as SME estimates can be imprecise.

Figure B.1
III Marine Expeditionary Force Hours Worked per Storage Marine per Day, January 2013–May 2018 (Preliminary Model)

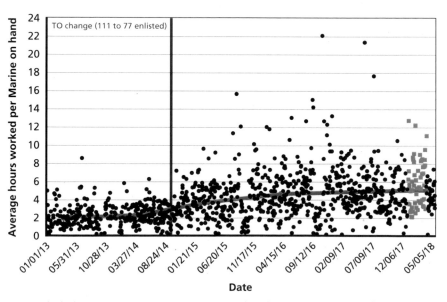

NOTE: Black dots represent average hours calculated using assumption of 42 percent of personnel on hand; green dots represent accurate on-hand data provided by III MEF.

Model Validation

Both the preliminary model described in the previous section and the final model presented in Chapter Three are based on averages. For each day in the time period considered, the number of times a task was performed was multiplied by the average task time, all task times were totaled, and, except for the data points in green, the total time was divided by an average number of storage Marines on hand during the III MEF time study. Under the circumstances, these models do not account for variance in task times or on-hand personnel.

To validate the models while allowing for variance in task times, we performed a Monte Carlo simulation, which randomly samples task times according to a normal distribution about the mean and allows for variation in the number of Marines on hand. Task times for which we had standard deviations are listed in Table 3.1. (The standard deviations were calculated from task-time samples provided in the III MEF Capacity Study.) In cases where we did not have standard deviations, we continued to treat the task times as averages. The simulation steps are described below.

Group 1 Tasks

1. Randomly select the number of stows that need to be performed from the III MEF FY 2018 STRATIS data set.
2. For each stow, randomly sample a receiving time from a normal distribution with mean and standard deviation listed in Table 3.1. Sum all receiving times and add to total time.
3. Add average care and storage time to total time.
4. For 5 percent of stows, add average causative research time to total time.
5. For 25 percent of stows, add average MHE operations time to total time.
6. For 25 items, add average quality control time to total time.
7. For each stow, randomly sample a stowing time from a normal distribution with mean and standard deviation listed in Table 3.1. Sum all stow times and add to total time.

8. For 25 percent of stows, add average MHE operations time to total time.
9. For 5 items, add average quality control time to total time.

Group 2 Tasks

10. Randomly select the number of picks that need to be performed from the III MEF FY 2018 STRATIS data set.
11. For each pick, randomly sample a picking time from a normal distribution with mean and standard deviation listed in Table 3.1. Sum all picking times and add to total time.
12. For 25 percent of picks, add average MHE operations time to total time.
13. For 5 items, add average quality control time to total time.
14. For each pick, randomly sample a packing time from a normal distribution with mean and standard deviation listed in Table 3.1. Sum all packing times and add to total time.
15. For 25 percent of picks, add average quality control time to total time.
16. For 25 percent of picks, add average MHE operations time to total time.
17. For each pick, randomly sample a shipping time from a normal distribution with mean and standard deviation listed in Table 3.1. Sum all shipping times and add to total time.
18. For 40 percent of picks, add average quality control time to total time.

Group 3 Tasks

19. Randomly select the number of inventories that need to be performed from the III MEF FY 2018 STRATIS data set.
20. For each inventory, randomly sample an inventory time from a normal distribution with mean and standard deviation listed in Table 3.1. Sum all inventory times and add to total time.
21. Add average causative research time to total time.
22. Add average rewarehouse time to total time.
23. Add average change location time to total time.

Calculate Average Hours Worked

24. Randomly select the number of storage Marines on hand from the 14-week-long III MEF Capacity Study.
25. Divide total time (in hours) by the number of Marines on hand.
26. Repeat steps 1 through 25.

We ran the steps above 1,460 times (covering four years, excluding weekends and holidays). A histogram of the results is shown below in Figure B.2. The average number of hours worked per day per Marine on hand on storage tasks, as estimated by the simulation, was approximately 7.7 hours (with a median of about 6.9 hours). These values are similar to the findings in Chapter Three and support the notion that storage Marines are currently working more than the ideal five to six hours per day on storage tasks.

Figure B.2
Histogram of Simulation Results

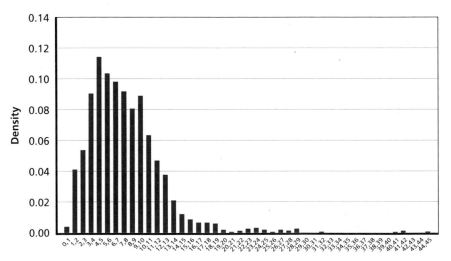

References

1st Supply Battalion, *Supply Management Unit Standard Operating Procedures,* SMUP 4400.1, November 16, 2017,

Abercrombie, Erick, Patrick Fullbright, and Seth Long, "Analysis of the Marine Corps Supply Management Unit's Internal Operations and Effect on the Warfighter," MBA Professional Report, Monterey, Calif.: Naval Postgraduate School, 2016. As of October 10, 2018:
http://www.dtic.mil/dtic/tr/fulltext/u2/1030643.pdf.

Broyles, James R., Shawn McKay, Albert A. Robbert, Kristin Van Abel, Maria DeYoreo, Cedric Kenney, and Kristin J. Leuschner, *Staffing Models for Customs and Border Protection's Support Services: A Generic Methodology and Specific Applications*, Santa Monica, Calif.: RAND Corporation, RR-2553-DHS, forthcoming.

Burke, Crispin J., "No Time, Literally, for All Requirements," Association of the United States Army, 2016. As of October 8, 2018:
https://www.ausa.org/articles/no-time-literally-all-requirements

Combat Logistics Regiment 35, 3rd Marine Logistics Group, "Projecting Combat Service Support Through Capacity, Utilization, & Integration," information brief, October 27, 2017.

"Ergonomics: The Backbone of a More Productive Warehouse," Ergonomics Supplement, *Modern Materials Handling*, April 2004, p. S4. As of October 8: 2018:
http://www.mhi.org/media/members/14599/129355195309038597.pdf.

Feickert, Andrew, "Marine Corps Drawdown, Force Structure Initiatives, and Roles and Missions: Background and Issues for Congress," Washington, D.C.: Congressional Research Service, CRS Report No. R43355, 2014. As of October 10, 2018:
https://fas.org/sgp/crs/natsec/R43355.pdf

Forrester, Jay W., *Industrial Dynamics*, Cambridge, Mass.: MIT Press, 1961.

Hemler, Joslyn, Yuna H. Wong, Walter L. Perry, and Austin Lewis, *Developing a Capacity Assessment Framework for Marine Logistics Groups*, Santa Monica, Calif.: RAND Corporation, RR-1572-USMC, 2017. As of October 11, 2018: https://www.rand.org/pubs/research_reports/RR1572.html

International Association for Six Sigma Certification, "Green Belt Certification," 2018a. As of October 12, 2018: https://www.iassc.org/six-sigma-certification/green-belt-certification/

———, "Yellow Belt Certification," 2018b. As of October 12, 2018: https://www.iassc.org/six-sigma-certification/yellow-belt-certification/

Nataraj, Shanthi, Christopher Guo, Philip Hall-Partyka, Susan M. Gates, and Douglas Yeung, *Options for Department of Defense Total Workforce Supply and Demand Analysis: Potential Approaches and Available Data Sources*, Santa Monica, Calif.: RAND Corporation, RR-543-OSD, 2014. As of October 10, 2018: https://www.rand.org/pubs/research_reports/RR543.html

Priest, Raymond, "Doing More with Less," *Marine Corps Gazette*, Vol. 74, No. 10, 1990. As of October 10, 2018: https://www.mca-marines.org/gazette/1990/10/doing-more-less

Roy, Michael M., Nicholas J. S. Christenfeld, and Meghan Jones, "Actors, Observers, and the Estimation of Task Duration," *Quarterly Journal of Experimental Psychology*, Vol. 66, No. 1, 2013, pp. 121–137. As of October 17, 2018: http://journals.sagepub.com/doi/pdf/10.1080/17470218.2012.699973

Siddiqui, Rafay A., Frank May, and Ashwani Monga, "Reversals of Task Duration Estimates: Thinking How Rather than Why Shrinks Duration Estimates for Simple Tasks, but Elongates Estimates for Complex Tasks," *Journal of Experimental Social Psychology*, Vol. 50, January 2014, pp. 184–189.

U.S. Marine Corps, *MAGTF Supply Operations*, Marine Corps Tactical Publication 3-40H, Washington, D.C.: Headquarters Marine Corps, February 19, 1996, updated through May 2, 2016. As of October 11, 2018: https://www.marines.mil/Portals/59/Publications/MCTP%203-40H.pdf?ver= 2017-03-23-102358-587

———, Marine Corps Order P4400.151B, Ch 2, *Intermediate-Level Supply Management Policy Manual*, Washington, D.C.: Headquarters Marine Corps, December 14, 2012. As of October 17, 2018: https://www.marines.mil/Portals/59/Publications/MCO%20P4400.151B%20 W%20CH%201-2.pdf

———, Marine Corps Order 4400.201, *Management of Property in the Possession of the Marine Corps*, Vol. 3, Washington, D.C.: Headquarters Marine Corps, June 13, 2016.

————, *Marine Corps Operating Concept: How an Expeditionary Force Operates in the 21st Century*, Washington, D.C.: Headquarters Marine Corps, September 2016. As of July 5, 2019:
https://www.mccdc.marines.mil/Portals/172/Docs/MCCDC/young/
MCCDC-YH/document/final/Marine%20Corps%20Operating%20Concept%20
Sept%202016.pdf?ver=2016-09-28-083439-483

————, "Types of MAGTFs," *Marine Corps Concepts and Programs*, Washington, D.C.: Headquarters Marine Corps, 2017. As of October 8, 2018:
http://www.candp.marines.mil/Organization/MAGTF/Types-of-MAGTFs/

Wilson, J. R., "Marine Corps Update: The Frugal Force Faces More Cuts," *Defense Media Network*, 2014. As of October 8, 2018:
https://www.defensemedianetwork.com/stories/
marine-corps-update-the-frugal-forces-faces-more-cuts

Wulfraat, Marc, "5 Ways to Improve Order Picking Productivity," *Warehouse/DC News*, May 15, 2013. As of November 29, 2018:
https://www.supplychain247.com/article/5_ways_to_improve_order_picking_
productivity/MWPVL_International